Best Practices for Teaching Statistics and Research Methods in the Behavioral Sciences

Best Practices for Teaching Statistics and Research Methods in the Behavioral Sciences

Edited by

Dana S. Dunn
Moravian College

Randolph A. Smith
Kennesaw State University

Bernard C. Beins
Ithaca College

LAWRENCE ERLBAUM ASSOCIATES, PUBLISHERS
2007 Ma' London

Lawrence Erlbaum Associates, Inc., Publishers
10 Industrial Avenue
Mahwah, New Jersey 07430
www.erlbaum.com

Cover design by Tomai Maridou

CIP information for this volume may be obtained by contacting the Library of Congress

p. cm.
Includes bibliographical references and index.
ISBN 0-8058-5747-8 (cloth : alk. paper)
ISBN 0-8058-5746-X (pbk. : alk. paper)
ISBN 1-4106-1464-6 (ebook)

Books published by Lawrence Erlbaum Associates are printed on acid-free paper, and their bindings are chosen for strength and durability.

Printed in the United States of America
10 9 8 7 6 5 4 3 2 1

DEDICATION

For my indomitable mother, Dah Kennedy Dunn–DSD

For Corliss–I admire your spirit–RAS

For Simon, Agatha, and Linda, Always–BCB

Contents

Part III Approaches to Teaching Statistics

Part IV Emerging Approaches to Teaching Research Methods

Part V Integrative Approaches to Teaching Methods and Statistics

Foreword

William Buskist
Auburn University

Of all the many things that we attempt to teach our students, especially our majors, perhaps the most important is how to think like scientists. Learning how to think scientifically involves acquiring a set of cognitive and behavioral problem-solving skills that have applicability well beyond the confines of the psychology laboratory. To be sure, scientific thinking can apply to every facet of life, whether it is to relatively simple matters such as making a wise consumer choice or to more complex matters such as understanding the subtleties of one's own interpersonal behaviors. Those of us who teach research methods and statistics courses understand perfectly the utility of scientific thinking and are excited, even "pumped," to share such insights with our students. As illogical as it may seem to our students, getting up in the morning and looking forward to teaching them how to design an experiment or conduct a statistical analysis is just something we enjoy doing. It is an integral part of our *raison d'etre* as teachers.

Although the introductory course offers students a glimpse of the vast expanse of psychology, and subsequent topical courses afford students the opportunity to explore the subject matter in more detail, only research methods and statistics courses give students the chance to experience rigorous and sustained exposure to the strategies and tactics of scientific thinking. Unfortunately, teaching students to think like scientists is not simple for several compelling reasons.

First, many students do not perceive that learning to think scientifically is inherently interesting or fun. These same students, at least at the outset of the course, fail to see how scientific thinking relates in any meaningful way to their lives outside of the classroom. Imagine that! Given the choice between taking a statistics course and almost any other course in the undergraduate psychology curriculum, many students would prefer another course. Some of these students may even prefer to watch paint dry rather than take statistics.

Second, many students may fear the course because of (a) the workload, (b) the complexity of the material, (c) the math involved, (d) their lack of preparedness for the rigors of scientific thinking, or (e) some combination of these factors. As a result, students often put the research methods and statistics courses off for as long as possible. I've known many a graduating senior to take both courses in the final semester of their college careers.

Third, learning to think scientifically is plain hard work. It taxes the brain in ways that many other courses do not. After all, writing a term paper on bipolar disorder or the Big Five theory of personality is not quite as complex for most students as designing an experiment, conducting it, and writing it up in proper APA style. Tests in research methods and statistics courses are seldom multiple choice and short-answer essay but, rather, entirely problem-based and analytical in their structure.

Finally, for many students scientific thinking is an acquired taste-indeed, some students, despite our best efforts, will never acquire this particular taste. Unlike learning to drink alcohol or to smoke cigarettes, learning to think scientifically is not strongly influenced by peer pressure. It is a decidedly lonelier venture, for which a good teacher and good teaching may tip the balance in inspiring students to work diligently to master the nuances of learning how to think critically.

These impediments to many of our students' willingness to become scientific thinkers represent the key challenges that face us as teachers of research methods and statistics at the beginning of each and every new academic term–an opportunity for perpetual renewal in our critical thinking regarding our teaching of critical thinking. Teaching these courses is not simply a matter of transmitting content to our students and getting them to go through the motions to solve a few problems. Rather, these courses require us to understand and confront the contexts that our students bring with them to our classes. The courses require us to create stimulating, even fun, ways to teach them. In short, these courses require us to identify and break down the barriers that stand in the way of our students becoming critical thinkers and in our way of becoming better teachers.

To assist psychology teachers in more effectively teaching research methods and statistics in terms of both content and context, the Society for the Teaching of Psychology teamed up with the National Institute on the Teaching of Psychology and Kennesaw State University's Center for Excellence in Teaching and Learning to sponsor the Best Practices in the Teaching of Research Methods and Statistics conference in the early fall of 2004. This "BP" conference was a watershed event that marked the first time ever that psychologists gathered together to examine the nature, scope, and processes involved in teaching these particular courses. Like the two previous BP conferences, this conference yielded a treasure trove of insight, tips, and advice that warranted a more permanent record than conferences typically allow. Fortunately, Dana Dunn, Randy

Smith, and Barney Beins understood the importance of carrying the conference's lessons to a larger audience and have painstakingly edited the book you are now reading. This book represents cutting-edge approaches to teaching research methods and statistics courses in psychology and offers wise counsel in how we might improve science instruction–in terms of both content and context–within our discipline.

A careful reading of this volume will no doubt widen our perspectives on what we might accomplish as teachers of these courses. It will help us address the challenges faced by our students and to create more stimulating, supportive, and comfortable learning environments. Thus, reading this book and heeding its suggestions will help us help our students become better learners, and as a direct result, more effective critical thinkers. And that is our ultimate goal as teachers of these courses–to teach students to think scientifically.

Preface

Statistics and research methods courses are arguably the mortar that holds the behavioral sciences curriculum together. Pedagogically, the material taught in statistics and methods bridges the broad expanse of introductory offerings and the depth and necessarily narrower focus of advanced topics. These two courses, or increasingly, integrated or sequenced courses, contain the theoretical and empirical tools that help students—budding, future researchers—develop experiments, test theories, and expand knowledge while carefully controlling against bias and error in drawing inferences. Of course, students must learn the tools of the trade; thus, the role of teachers is of paramount importance. Students need to learn to wield statistical and methodological tools in their research, just as they need to understand the logic underlying the use of these tools, a logic that will help them understand upper level offerings in the behavioral science curriculum.

Statistics and research methods courses are also watershed courses for many students majoring in the behavioral sciences. Unfortunately, some students approach these courses with fear and trepidation, if not outright dread. Instead of thinking about the content of statistics and methods as helpful for conducting or interpreting research, all too often students view the courses as something to be endured and passed. Students often treat the detection of differences, the discovery of relations among variables through data analysis, and the development of critical empirical investigations through rigorous research designs more as a trial by fire than as essential knowledge. Teachers of these courses worry that students will not retain what they need to know; teachers of the courses that follow often complain that the transfer of knowledge is limited.

To remedy these problems, we designed our volume to offer good news and good reading. This collection of chapters does not simply provide solace for behavioral science teachers, but also practical advice and practices that will make statistics and research methods courses memorable for students. More to the point, the best practices detailed herein will help teachers improve what and how they teach in these critically important courses, while offering a variety of helpful suggestions, exercises, and perspectives. Our authors represent the voice of experience and dedication, as each one has spent considerable time, energy, and effort in mastering a particular approach to effectively teaching statistics, methods, or a combination of both.

This book is the result of the *Finding Out: Best Practices for Teaching Statistics and Research Methods in Psychology Conference*, which was held in Atlanta, Georgia, October 1–2, 2004, with the support of Kennesaw State University. During the 2-day event, over 250 participants attended a variety of sessions devoted to improving the delivery of statistics and research methods in the behavioral sciences classroom. The conference was sponsored by the Society for the Teaching of Psychology (STP), the Kennesaw State Center for Excellence in Teaching and Learning (CETL), and the National Institute on the Teaching of Psychology (NITOP). In the months following the conference, we gathered a group of nationally recognized teachers to craft a book portraying the current best pedagogical practices for presenting the material found in statistics and research methods courses in the behavioral sciences.

With our authors' guidance and skills, we offer postsecondary educators who teach the critical intermediate courses of statistics and research methods sage advice on:

Involving undergraduate students in research and teaching them how best to do it. How can teachers help students develop researchable project ideas?

Current perspectives on the changes and developments in the teaching of statistics and *Developing intermediate, advanced, or online courses.* How can teachers develop an online statistics course? How do teachers go beyond the basics in statistics?

Creating hands-on lab experiences for students. How can teachers integrate research methods concepts in nonmethods courses? How can teachers get through the material and develop student projects in a single semester?

Adding computer applications to statistics or methods courses. Can teachers develop online courses to deliver this type of material?

Attending to assessment issues. How can teachers effectively assess student learning in statistics and research methods courses?

Linking diversity to research methodology. Are research methods courses devoid of opportunities to integrate diversity?

Turning statistics and research methods courses into writing-intensive opportunities. Is there a place for writing in these courses?

ACKNOWLEDGMENTS

The development, execution, and production of this book went smoothly because of the good work of our authors and the professionalism of the Lawrence Erlbaum Associates editorial and production team. At Erlbaum, we thank our editor, Debra Riegert, for her encouragement and vision from the start of the project. We appreciate the organizational efforts of Kerry Breen, Rebecca Larsen, Suzanne Sheppard, Sarah Wright and the LEA production staff. We are especially grateful to those colleagues who pulled the *Finding Out* conference together: Bill Hill, Jane Halonen, Doug Bernstein, Maureen McCarthy, Randy Smith; the Psychology faculty and CETL staff at Kennesaw State University; and Linda Noble. We thank our colleague, Michelle E. Schmidt, for organizing the chapter materials on the CD that accompanies our book. As usual, Dana Dunn relied on the reliable efforts of Jackie Giaquinto to finish the project on time and is also grateful to Moravian's Faculty Development and Research Committee for providing him with travel funds and a 2004 Summer Stipend to further work on the book. Randolph Smith is appreciative of Kennesaw State University's support of this conference and Bill Hill's tutelage in helping to design and run such conferences. Bernard Beins appreciates Ithaca College's promotion of a vibrant teaching environment and the support it has provided for the scholarship of teaching.

We thank our peer reviewers who made helpful suggestions on our project proposal and who reviewed and constructively commented on the proposal for this book: Eric Landrum (Boise State University), Kevin J. Apple (James Madison University), Douglas A. Bernstein (University of South Florida), and Kenneth Weaver (Emporia State University).

In the end, of course, we thank our families who, though long used to our busy schedules, were especially understanding during the various phases of this project. We thank them for their love and patience.

Dana S. Dunn
Randolph A. Smith
Bernard C. Beins

Contributors

Ruth L. Ault, Davidson College–ruault@davidson.edu
Kenneth E. Barron, James Madison University–barronke@jmu.edu
Cole Barton, Davidson College–cobarton@davidson.edu
Bernard C. Beins, Ithaca College–beins@ithaca.edu
James O. Benedict, James Madison University–benedijo@jmu.edu
Dolores V. Bradley, Spelman College–dbradley@spelman.edu
Karen Brakke, Spelman College–kbrakke@spelman.edu
William F. Buskist, Auburn University–buskiwf@auburn.edu
Arlene R. Casiple, James Madison University–casiplar@jmu.edu
Stephen L. Chew, Samford University–slchew@samford.edu
Andrew N. Christopher, Albion College–achristopher@albion.edu
Stephen F. Davis, Texas Wesleyan University, Emeritus–davis122@cox.net
Dana S. Dunn, Moravian College–dunn@moravian.edu
Haslan Hamdan, James Madison University–hamdanhx@jmu.edu
Charles M. Harris, James Madison University–harriscm@jmu.edu
G. William Hill, IV, Kennesaw State University–bhill@kennesaw.edu
Robert S. Horton, Wabash College–hortonr@wabash.edu
Michael R. Hulsizer, Webster University–hulsizermr@webster.edu
Edward P. Kardas, Southern Arkansas University–epkardas@saumag.edu
Kenneth D. Keith, University of San Diego–kkeith@sandiego.edu
Roger E. Kirk, Baylor University–Roger_Kirk@baylor.edu
R. Eric Landrum, Boise State University–elandru@boisestate.edu
Ann Lynn, Ithaca College–alynn@ithaca.edu

Pam Marek, Kennesaw State University–pmarek@kennesaw.edu

James Mazou, James Madison University–mazouejg@jmu.edu

Jody Meerdink, Nebraska Wesleyan University–jem@nebrwesleyan.edu

Adriana Molitor, University of San Diego–amolitor@sandiego.edu

Kristi S. Multhaup, Davidson College–krmulthaup@davidson.edu

Margaret P. Munger, Davidson College–mamunger@davidson.edu

Bryan K. Saville, James Madison University–savillbk@jmu.edu

Michelle E. Schmidt, Moravian College–mschmidt@moravian.edu

Sherry L. Serdikoff, James Madison University–serdiksl@jmu.edu

Paul C. Smith, Alverno College–paul.smith@alverno.edu

Randolph A. Smith, Kennesaw State University–rsmith@kennesaw.edu

Chris Spatz, Hendrix College–spatz@hendrix.edu

Scott Tonidandel, Davidson College–sctonidandel@davidson.edu

Mark I. Walter, Albion College–mwalter@albion.edu

Janie H. Wilson, Georgia Southern University–JHWilson@georgiasouthern.edu

Linda M. Woolf, Webster University–woolflm@webster.edu

Tracy E. Zinn, James Madison University–zinnte@jmu.edu

Part I

Introducing Best Practices

Chapter 1

Overview: Best Practices for Teaching Statistics and Research Methods in the Behavioral Sciences

Dana S. Dunn
Moravian College

Randolph A. Smith
Kennesaw State University

Bernard C. Beins
Ithaca College

Scholarship on teaching about research methods and statistics effectively in the behavioral sciences is sparse. This state of affairs is ironic because most psychologists recognize that the minimum requirements for a quality undergraduate education in psychology include teaching students how to develop hypotheses; conceive research projects; collect and analyze data using statistics; and to then summarize, write reports, and present research results. Most psychology programs—whether based in a liberal arts college or a research university—want their graduates to know how to "find out" what factors cause behavior to occur, an expectation that can only be met when students apply the tools and techniques of methodology and statistics. In fact, beyond introductory psychology, among the most common "core courses" found in psychology are apt to be research methods and statistics for psychology or some offering designed to integrate the two (e.g., Brewer et al., 1993; Perlman & McCann, 1999, 2005).

Instructors who teach methods and statistics must confront the fact that many students view these required courses with a mix of emotions, most of them not very favorable. Students fear that these courses will be overly dull or, worse yet, much too challenging for them to succeed. When students complain about courses in the psychology curriculum, research methods and statistics receive the brunt of the criticism. The fact is that the courses have become more demanding over time—many, perhaps most, methods and statistics courses require computing skills; others have a hefty writing requirement (on many campuses, methods and statistics are designated as "writing intensive" courses).

Novice teachers who have the responsibility of teaching methods or statistics often end up feeling similar to their students in their uncertainty about the courses. They worry about how to present the course material in an engaging but rigorous manner that is inviting to the students. Candidly, many instructors also worry about their course evaluations. More than a few teachers leave graduate school to discover that, as the junior members of psychology departments, they must assume responsibility for one (and occasionally both) of the most difficult offerings. Of course, veteran psychology instructors are also not immune to these challenges. They are likely to list methods and statistics as the most difficult psychology courses to teach well (along with introductory psychology), and instructors continue to grapple with the challenges of this course throughout their career.

Teaching research methods and statistics well is essential because psychology majors must be able to evaluate critically the psychological research they encounter in upper level courses. For effective teaching, they must have a firm grounding in the scientific method; experimental, quasi-experimental, and qualitative research methods; research ethics (including informed consent and debriefing); measurement and manipulation of variables; formulating and testing hypotheses; collecting, coding, and analyzing data; and writing up results in accepted, scientific style (i.e., American Psychological Association [APA] style). Some psychology majors will continue their education at the graduate level, which means that their acquired methods and statistical skills will help them eventually contribute to the discipline by conducting and sharing scientific research.

Further, all students must learn how to distinguish scientific psychology from popular and pseudoscientific views. Skills that students develop in statistics and research methods are useful for understanding this important distinction. In addition, the mode of scientific thought engendered in statistics and research methods aids students pursuing careers in a variety of areas, not just for graduate school.

Our book is a scholarly but pedagogically practical attempt to address concerns regarding teaching research methods and statistics. We provide a showcase for best practices in teaching statistics and research methods in 4-year colleges and universities, 2-year colleges, and high schools. Our authors give readers more than mere teaching tips by emphasizing a scholarly approach to teaching these related courses. We asked our authors to be certain to address two specific topics in their chapters: a practical orientation and assessment issues.

PRACTICAL ORIENTATIONS FOR BEST PRACTICES

In developing their contributions, our authors focused on the practical, that is, on current pedagogy that works in the classroom (whether face-to-face or online) and did so in new ways. Indeed, we gave them a charge of identifying ways to teach students to learn effectively, especially because the material concerning data analysis and methodology is often abstract, unfamiliar, and highly technical. Because of these abstractions and complexities, students often enroll in methods and statistics courses with the expectation that the work will be very challenging. By adopting a practical perspective, our authors were determined to provide teachers with tools and activities designed to help their students learn the material presented in either or both classes. Consider one simple example: Research methods and statistics courses tend to be lecture-based, but our authors suggest ways that teachers can actively involve students in meaningful classroom discussions and activities about the course materials.

Similarly, when it comes to innovative teaching techniques, most teachers of statistics and research methods are left to their own devices or they must rely on the Instructor's Manual (IM) accompanying their chosen textbook (if one is even available). The IMs accompanying methods and statistics books usually contain few pedagogical tools or suggestions for classroom activities beyond coverage of the material; in fact, these ancillaries are often little more than glorified test banks. In contrast, our contributors share novel pedagogical resources available and, where possible, discuss empirical findings concerning those that are most educationally effective.

To go beyond their written descriptions of what they do in the classroom, we invited our authors to submit supplementary materials for our readers. Compiled by our colleague, Michelle E. Schmidt of Moravian College, the CD accompanying the book contains a variety of teaching materials–everything from sample syllabi to writing assignments–created by many of our chapter authors. We believe these additional resources will enable teachers of statistics and methods to revise or refine their teaching approaches with confidence.

The Importance of Assessment

Assessment is a key watchword in pedagogy in psychology (Dunn, Mehrotra, & Halonen, 2004; Halonen et al., 2003). *Educational assessment*–the measurement and evaluation of whether students learn what we teach them–is especially important in statistics and research methods courses. Do most courses covering either topic live up to their educational goals? How do we know students have learned to adequately analyze data, for example? Can we be certain that the conceptual and practical elegance of research design is carried forward for application in other courses or in actual research efforts undertaken by students? We challenged our authors to reflect on what role assessment already plays or should play in their domain of expertise and interest. Thus, the chapters in this book offer assessment advice to teachers and administrators so they can demonstrate that student learning, the transfer of reasoning ability from the classroom to the lab or field, is indeed occurring following methods and statistics classes.

BEST PRACTICES FOR TEACHING STATISTICS AND RESEARCH METHODS

Basic Issues for Best Practices

Steven Davis (chap. 2) discusses the pedagogical importance of establishing collaborative relationships between students and faculty members doing shared research. Reflecting on his long and distinguished teaching career, Davis describes how working closely with students on research can benefit both the learners and the teacher. He describes several collaborative models that allow for productive, intellectual experiences for the participants. His preference is for what he calls the "junior colleague model," in which the students identify research ideas and continually improve the process involved in exploring the topics. Davis and his student colleagues conducted research on a year-round basis, a choice that allows for flexibility, reflection, and an influx of new junior colleagues at different points in time. Readers will learn that students can produce solid research with minimal funds and that there are decided benefits to collaboration for both students and faculty colleagues.

A renowned statistician and teacher, Roger Kirk (chap. 3) discusses the surprising growth and change found in the content of undergraduate statistics courses during the last few decades. As Kirk explains, the ubiquity of computers, statistical software, and even handheld calculators are only part of the story of pedagogical development in statistics education. The introduction of new topics, coupled with emerging techniques, indicate that today's undergraduate students are routinely expected to know how to address problems that, once upon a time not so long ago, were the purview of graduate students in statistics or quantitative methods. Kirk concisely lays out the history of progress in statistics pedagogy, including thoughtful comments on (a) the controversy surrounding null hypothesis testing, (b) the virtues of computing effect size and confidence intervals, and (c) the dramatic increase in the number of students now taking courses in data analysis. Kirk closes his thoughtful chapter by making forecasts about the future in statistical pedagogy.

Eric Landrum and Randolph Smith (chap. 4) discuss considerations that faculty should make in developing syllabi. They briefly review literature concerning constructing good syllabi but focus specifically on syllabi for statistics and research methods courses. They derive their specific suggestions based on both the published literature and syllabi that they collected and analyzed. For both courses, Landrum and Smith cover course objectives and textbooks. For statistics syllabi, they add the topics of course organization and use of statistical software and calculators. For methods syllabi, they add the topics of research projects, research ethics, and APA format.

In chapter 5, Paul Smith identifies some of the pitfalls associated with helping students create experimental research projects. Among the difficulties is recognizing the difference between true experimental designs with manipulated independent variables and quasi-experimental designs using preexisting categories. Further, students often fall prey to the desire to conduct a study that makes a fundamental breakthrough in understanding psychological thought, a plan that would fail even for experienced researchers. On the other end of the continuum is the problem of creating experiments that are relatively trivial and that do not add measurably to knowledge of psychological processes. Smith also notes that by attending to student problems, an instructor can identify important elements of students' understanding of psychology and of the nature of the research process. By becoming aware of student views, the instructor can provide useful guidance for the development of the research idea.

Approaches to Teaching Statistics

In chapter 6, Stephen Chew discusses using effective examples and problems in teaching statistics. Chew reviews the literature about how to create and use examples effectively, particularly in a statistics course. He shows the properties of good examples and what students are likely to learn from examples. Chew ends his chapter by proposing a model of how to design and use examples effectively that should help statistics teachers create good examples more systematically.

The next chapter (7), by James Madison University colleagues Charles Harris, James Mazoué, Hasan Hamdan, and Arlene Casiple provides an overview of the process of conceptualizing online courses and then actually constructing them. An online course can reflect different conceptual frameworks, most notably those of Piaget or of Vygotsky. Once the theoretical framework is established, more practical matters of course software and technical considerations arise. The authors also discuss seven principles for good pedagogy relating to faculty–student interactions, student–student interactions, active learning, assessment, time on task, setting high expectations, and respecting diverse ways of learning.

In chapter 8, Karen Brakke, Janie Wilson, and Dolores Bradley reflect on how best to include advances in statistics into the undergraduate classroom. These advances including evolving standards and practices, multivariate analyses, and complements to null hypothesis testing. The authors present two ways for instructors to augment the basic statistics curriculum offered in most behavioral science departments. Their first approach involves streamlining existing one-semester courses so that instructors can include additional material (e.g., effect size). Their second approach consists of developing a supplementary, upper level statistics course that builds on students' inferential facility with various advanced statistics. The strength of Brakke, Wilson, and Bradley's work is their focus on demonstrating how students understand and apply the advanced statistics taught in these updated courses.

Emerging Approaches to Teaching Research Methods

In chapter 9, colleagues from Davidson College—Ruth Ault, Margaret Munger, Scott Tonidandel, Cole Barton, and Kristi Multhaup—discuss how teachers can introduce hands-on laboratory activities into content area research methods courses. The Department of Psychology at Davidson is fortunate to offer 11 different research methods courses. This chapter discusses lab activities for methods courses in perception and attention, industrial/organizational psychology, clinical psychology, and adult development. The different activities reflect the myriad ways that psychologists ask and answer questions using empirical methods. The advantage for students who are fortunate to encounter such content area methods courses is having their learning grounded in a professional, scientifically-based approach to doing psychological research. Beyond covering the strengths and weaknesses of approaches to inquiry in each area, the Davidson colleagues describe ways to enhance students' abilities to read primary sources, to design experiments, and to collect, analyze, and present research findings.

Kenneth Barron, James Benedict, Bryan Saville, Sherry Serdikoff, and Tracy Zinn, from James Madison University, present three innovative ways of approaching the research methods course in chapter 10. In *Just-in-Time Teaching*, students answer questions posted by the instructor on the Web just before class. In this manner, the instructor has a good sense of what students do and do not understand immediately before class. *Interteaching* entails giving students a prep guide to help them read material before the day of coverage in class. During the actual interteaching session, small groups of students discuss the reading material and the prep guide. At the end of the discussion, students complete a form that informs the instructor about any points of uncertainty; the instructor uses these forms to structure a short lecture at the beginning of the next class. The James Madison faculty has developed a *learning community* for the statistics and research methods classes. They combined the statistics and research methods information over two semesters and taught an intact group from the residence halls. Preliminary data show a great increase in the percentage of students who conduct independent research projects from the communities.

Edward Kardas and Chris Spatz contribute important ideas to the teaching of ethics in research (chap. 11). Ethical standards, as they point out, arise from a combination of longstanding ideas of Western culture combined with specific principles of ethical issues related to empirical research. They discuss important cases of psychological and other scientific research that involve serious ethical considerations. Kardas and Spatz also highlight the ethical standards that the American Psychological Association has developed for researchers. They conclude with a presentation of issues that affect students directly as they create their studies and submit them for institutional review board approval.

In chapter 12, Kenneth Keith (University of San Diego), Jody Meerdink (Nebraska Wesleyan University), and Adriana Molitor (University of San Diego) discuss strategies for having students work on a research project from start (getting the idea) to finish (completing the final write-up) in a single semester. Although all three are working with students in different venues (Keith: cross-cultural psychology laboratory; Meerdink: senior thesis; Molitor: developmental psychology laboratory), they extract common features from these experiences and provide valuable pointers for readers.

Integrative Approaches to Teaching Methods and Statistics

In the next chapter (13), Andrew Christopher and Mark Walter from Albion College, Robert Horton from Wabash College, and Pam Marek from Kennesaw State University identify benefits and detriments of integrating statistics and research methods in one course. This growing trend can proceed well only with a serious consideration of the best way to guarantee that students will understand both the statistical and the design issues in research. As these authors point out, combining these heretofore separate courses provides the context for the two related domains and mimics the activity of researchers. In their chapter, the authors also note the pragmatics of student research, including such factors as managing course time, using statistical software, and tracking student progress from ideas to execution of the research. Finally, the authors discuss issues that departments must face in combining these courses.

Ithaca College colleagues Bernard Beins and Ann Lynn have identified issues associated with incorporating computers into statistics and research methods courses (chap. 14). They note that it is critical for teachers to adopt appropriate technology and to balance conceptual aspects of statistics and research with the technical aspects. Further, the authors discuss the balance between traditional (and successful) teaching modes and newer technologically based approaches. Beins and Lynn also pose questions about effective use of the Internet in presenting material, collecting data, presenting homework, and administering tests. Finally, the authors identify potential ways of assessing student outcomes associated with the approaches they discuss.

In chapter 15, Randolph Smith focuses on providing professional opportunities for students who have gone through the entire research process and have completed a study and its write-up. Just as professional psychologists attend conferences to give research talks and submit manuscripts for publication, Smith contends that undergraduates should have the same opportunities in order to experience the full range of professional opportunities and obligations.

Special Topics: Diversity Issues and Writing

We conclude the book by turning to two special topics–diversity and writing–that often receive too little attention in the hectic pace typically found in the teaching of statistics and methods. Linda Woolf and Michael Hulsizer (chap.16) tackle a neglected problem, the infusion of diversity issues into research methods courses. Increasingly, psychology curricula are becoming relatively more inclusive by attending to issues of race, gender, culture, and ethnicity, among other topics portraying the experience of diverse groups. Perhaps the presumed objectivity of research methods courses explains the dearth of diverse content within them. Woolf and Hulsizer counter that ample material concerning methodology and diversity exist—instructors need only tap into some of the available resources in order to better represent human experience by updating course content. Although these authors examine six areas of diversity (race/ethnicity, gender, sexual orientation, age, disability, cross-cultural/international), they emphasize that these areas are not mutually exclusive, nor do these areas exhaust diverse possibilities. Once readers are armed with Woolf and Hulsizer's topical examples as templates, they can search the psychological literature for other examples of human diversity for teaching research methodology.

Finally, in the book's last chapter (chap.17), Michelle Schmidt and Dana Dunn tackle the all-important role of writing, especially the use of APA style, in statistics and research methods courses. Schmidt and Dunn describe the writing intensive requirement used in Moravian College's two-semester, integrated statistics and methods courses, Experimental Methods and Data Analysis I and II. Besides gradually teaching students to craft an APA style research report from conception to completion, these colleagues also discuss how to use a variety of graded and ungraded exercises to teach writing in psychology. The chapter also presents a variety of non-APA style activities that focus on improving students' general communication and writing skills.

CLOSING THOUGHTS

The myriad of thoughtful and clever approaches to developing and renewing statistics and research methods courses in this volume bespeak the energy and enthusiasm of the psychologists who teach these courses. The chapters reflect the ways the authors have grappled with complex, conceptual issues, along with the practical decisions they have made in creating their classes. The result is an information compendium that will assist new teachers to develop their pedagogical styles and philosophies and will also engage experienced teachers.

ACKNOWLEDGMENTS

We are grateful to G. William Hill, IV and his colleagues at the Center for Excellence in Teaching and Learning at Kennesaw State University.

REFERENCES

Brewer, C. L., Hopkins, J. R., Kimble, G. A., Matlin, M. W., McCann, L. I., McNeil, O. V., Nodine, B. F., Quinn, V. N., & Saundra. (1993). Curriculum. In T. V. McGovern (Ed.), *Handbook for enhancing undergraduate education in psychology* (pp. 161–182). Washington, DC: American Psychological Association.

Dunn, D. S., Mehrotra, C., & Halonen, J. S. (Eds.). (2004). *Measuring up: Educational assessment challenges and practices for psychology*. Washington, DC: American Psychological Association.

Halonen, J. S., Bosack, T., Clay, S., & McCarthy, M. (with Dunn, D. S., Hill, G. W., IV, McEntarfer, R., Mehrotra, C., Nesmith, R., Weaver, K., & Whitlock, K.). (2003). A rubric for authentically learning, teaching, and assessing scientific reasoning in psychology. *Teaching of Psychology, 30*, 196–208.

Perlman, B., & McCann, L. I. (1999). The structure of the psychology undergraduate curriculum. *Teaching of Psychology, 26*, 171–176.

Perlman, B., & McCann, L. I. (2005). Undergraduate research experiences in psychology: A national study of courses and curricula. *Teaching of Psychology, 32*, 5–14.

Part II

Basic Issues for Best Practices

Chapter 2

Student–Faculty Research Collaboration: A Model for Success in Psychology

Stephen F. Davis
Texas Wesleyan University

During my graduate-school years, I unquestioningly assumed that my future would revolve around teaching and conducting research at a college or university. I also assumed that, just as I was included in research projects as a student, students also would be an integral component of my research activities. Although these assumptions proved to be accurate, the involvement of students in my research projects took an unanticipated turn.

Graduate training in basic animal learning processes and comparative psychology led me to establish animal research laboratories during my professional career at King College (Bristol, TN), Austin Peay State University (APSU; Clarksville, TN), and Emporia State University (ESU; Emporia, KS). The novelty of each of these new research labs attracted students and collaborative projects soon were under way. During the King College years, our research was focused exclusively on olfactory communication in animal maze learning and I am sure that my dissertation director was confident that I understood the value of conducting programmatic research. However, when I began teaching at APSU, a larger and more diverse institution, I soon realized that not all students shared my passion for studying olfactory communication in rats and mice. Many of the students were coming to my office to discuss proposed research topics that were very unfamiliar, at least to a person trained in the Hull-Spence tradition!

Thankfully, I had the presence of mind to give these unorthodox research ideas a fair hearing; we began to follow through on some of them. For example, projects on such topics as personality characteristics of civilian and military policemen, an analysis of the size of human figure drawing and level of self-esteem in school-age children, and death anxiety in military couples reached fruition. As we completed more of these atypical projects, I found that they had considerable interest and appeal to me. Yes, I found that my research focus was changing; programmatic research did not seem as important as it once had. I realized that my laboratory and research activities did not exist for the conduct of any specific type of research; they existed for the training of quality undergraduate students.

I also realized that my view of teaching had expanded considerably. Teaching about research methods and various data analytic techniques are the sum and substance of the classroom; however, teaching students about the full scope of the research process and the inherent enjoyment to be gained from being a part of this process required direct involvement. It is amazing how much teaching and learning takes place outside the classroom. These realizations bring us to the central theme of this chapter: student–faculty research collaboration.

SEVERAL MODELS OF STUDENT–FACULTY RESEARCH COLLABORATION

Students and faculty can engage in collaborative research ventures beyond the research methods/experimental psychology course in a variety of ways. I examine several of these collaborative models.

Single-Faculty, Single-Project Team

Gibson, Kahn, and Mathie (1996) described two models for involving undergraduate students in research groups used by faculty at their institution. In the first model, a single faculty member works with several students to conduct a single research project (typically of the faculty member's choosing). At the beginning of the project, the students list their current skills and the skills they would like to acquire as a result of working on the project. The faculty director and student researchers meet weekly to discuss the project's progress, plan for future steps, and complete any unfinished tasks. Gibson et al. (1996) indicated that student researchers who are members of research teams using this model at their institution are encouraged to present and publish their data.

You should keep in mind that, regardless of the model that faculty use, research collaboration is a two-way street; students bring the willingness to assist, learn, and contribute, whereas the faculty contributes expertise and guidance. However, both students and faculty must make meaningful contributions and avoid problems and unnecessary confrontations for this commitment to work effectively. Slattery and Park (2002) described several strategies that faculty can adopt to avoid problems in research collaboration with students: (a) Choose student researchers carefully, (b) have regular meetings with your student collaborators, (c) train your student collaborators carefully and thoroughly, and (d) serve as a mentor for your student collaborators.

Large Multifaculty, Multiproject Teams

The second model described by Gibson et al. (1996) involved three faculty members who, with several student researchers, simultaneously conducted four research projects on sexual assault. In order to accommodate additional student researchers, the structure of the large multifaculty, multiproject team may change and evolve. Although students affiliate with only one project group, they know the nature and progress of the projects that the other teams of student researchers are conducting. As with the single-faculty, single-team project model, the faculty supervisors hold weekly meetings that include all research teams. All student researchers have an opportunity to provide input into all projects at these meetings. Gibson et al. (1996) indicated that many student researchers who receive training under this model take advantage of opportunities to present and publish their data.

Research Internship/Assistantship/Practicum/Honors Project

The research internship (also classified as an assistantship or a practicum) offers students course credit (typically 1, 2, or 3 units) for participating as a research assistant. Effective uses of this model include the conduct of (a) a longitudinal, field-based project (Evans, Rintala, Guthrie, & Raines, 1981);, and (b) senior honors theses and independent study projects (Kierniesky, 1984; Starke, 1985). Although projects of this nature typically involve one student and one faculty supervisor, Dunn and Toedter (1991) reported an innovative, collaborative honors project that involved two students and two faculty advisors. Perhaps because academic credit is attached to the research internship/assistantship/practicum/honors project, there does not appear to be as strong a push for the student researchers to take their completed projects to the presentation and publication stages of development.

THE JUNIOR COLLEAGUE MODEL

I call the model I adopted the "junior colleague model" because the research ideas typically come from the students and because the students are always growing and developing in their ability to add meaningfully to a project(s). Under this model, we conducted research on a year-round basis; the inception and completion of projects were not necessarily linked to the beginning or end of the academic term. Some turnover in student researchers occurred as students graduated, projects were completed, or other commitments precluded participation in another project at that time. Moreover, I offered research students the opportunity to enroll for academic credit (1–3 hours as determined by the student and the supervisor) during a maximum of two semesters. Course enrollment was not obligatory and approximately 60% of my student researchers have taken advantage of the opportunity to obtain course credit. To my delight, the majority of the students who came to work on research were not overly concerned with accruing credit hours; it was my responsibility to provide them with the best experience possible.

Because we conducted research projects on a year-round basis, new students were able to join the "lab group" (as the students called it) at different times during the year. McKeachie's fine book, *Teaching Tips* (McKeachie, 2002), prompted my decision to implement peer teaching in the lab group; the more experienced student researchers acted as mentors to the neophytes. For example, entering data into an SPSS file might be the first assignment for a student working on a project using human participants. Similarly, neophytes working in the animal laboratory might begin their career by learning the techniques of good animal care. As students completed projects, they undertook new ones that involved increased involvement, input, and responsibility. I have found that my student researchers frequently were capable of making significant contributions to the design and implementation of an experiment by the time they were working on their second or third project.

During my years of working with my lab group I stressed two basic messages. You do not need expensive equipment and extensive financial support to conduct good research. What is important is that the quality of the research idea and student-generated ideas can be as good as the ideas of an established professional. Table 2.1 shows a sampling of some of the research ideas proposed by my students over the years. I believe my students, in addition to being highly diverse, have had some excellent research ideas over the years. I hope you agree.

BENEFITS FOR STUDENTS

In addition to the enjoyment the students have gained from conducting research, what other benefits can they expect to derive from this experience? Landrum and Nelsen (2002) surveyed undergraduate psychology educators in order to determine the benefits that supervising faculty believed their students accrued from their research involvement. Factor analysis of the data yielded two factors. Factor 1 dealt with preparation for graduate school and some employment opportunities available to undergraduate students once they graduate. The skills contributing to this factor included the ability to (a) use statistical programs and procedures to analyze data, (b) develop surveys, and (c) prepare a manuscript in accepted American Psychological Association (APA) format. Factor 2 contained such interpersonal skills items as leadership, self-confidence, teamwork, time management, and coping with deadlines. Although the importance of the two general factors reported by Landrum and Nelsen (2002) is undeniable, their generality may make them seem somewhat abstract. I have delineated several specific, attainable benefits students accrue from research involvement.

TABLE 2.1
Selected and Annotated Examples of Student-Generated Research

Davis, S. F., Grover, C. A., Becker, A. H., & McGregor, L. N. (1992). Academic dishonesty: Prevalence, determinants, techniques, and punishments. *Teaching of Psychology, 19*, 16–20.

This project, which illustrated a continuing interest in academic dishonesty, was prompted by Cathy Grover, witnessing another student engaging in cheating and expressing a desire to study this behavior.

Miller, H. R., & Davis, S. F. (1993). Recall of boxed material in textbooks. *Bulletin of the Psychonomic Society, 31*, 31–32.

Holly Miller wondered why authors and publishers were so fond of using boxed materials when students rarely, if ever, read these portions of the text. This experiment demonstrated that students do not read boxed materials unless the instructor explicitly tells them to do so.

Nash, S. M., Weaver, M. S., Cowen, C. L., Davis, S. F., & Tramill, J. L. (1984). Taste preference of the adult rat as a function of prenatal exposure to ethanol. *The Journal of General Psychology, 110*, 129–135.

Left over from a study of the fetal alcohol syndrome, we had two unique groups of adult rats: rats exposed to alcohol during gestation and litter mates that were water exposed during gestation. Susan Nash proposed a most elegant experiment: Determine the taste preference of these animals as adults. Her findings demonstrated that as adults, the animals exposed to alcohol during gestation preferred to drink ethanol over a 6-day period. This research won Susan the Psi Chi J. P. Guilford Undergraduate Research Award.

Tramill, J. L., Gustavson, K., Weaver, M. S., Moore, S. A., & Davis, S. F. (1983). Shock-elicited aggression as a function of acute and chronic e thanol challenges. *The Journal of General Psychology, 109*, 53–58.

As part of an ongoing project linking increases in aggressive behavior to various physiological states, James Tramill was interested in the effects of chronic exposure to ethanol. His results demonstrated that chronic exposure to alcohol results in increased levels of shock-elicited aggression in rats.

Admission to Graduate School

Research experience can be a crucial factor in determining a student's admission to graduate school (Landrum & Davis, 2004). For example, Keith-Spiegel, Tabachnick, and Spiegel (1994) surveyed members of doctoral selection committees. Based on the results of these surveys, Keith-Spiegel et al. were able to distinguish between first- and second-order admissions criteria. The primary or first-order, and presumably most important, criteria included grade point average, Graduate Record Examination scores, and letters of recommendation (see Keith-Spiegel & Wiederman, 2000). Because selection committees anticipate that the brightest and best students will apply for admission to graduate school, it is arguable that these committees actually use the primary criteria for initial screening purposes. Thus, selection committees actually may rely on the second-order criteria to determine which applicants will receive an invitation to join a specific program. According to Keith-Spiegel et al. (1994), the most important second-order criterion is research experience, especially research experience that leads to a convention presentation or journal presentation.

Thus, I also see the development of professional abilities and credentials, such as convention presentations and publications, as part of my junior colleague model responsibilities to my research students. This student professional development also extends to vita preparation and a discussion of career goals, whether they are job- or graduate-school oriented. In short, my "lab group" and the junior colleague model serve to train and develop students in the broadest possible ways.

Networking

Attending conventions to deliver oral papers or make poster presentations prompts consideration of another research-related benefit for students: networking. Without question, the ability to network effectively is a valuable professional skill, and one that students should be aware of and begin to cultivate early in their careers. I have encouraged my students to make two types of contacts when they attend conventions. First, I recommend that they talk with students from other institutions; many of today's students will be tomorrow's faculty members. Second, I believe that it is important for my students to talk with faculty from other institutions. From such faculty conversations, students can acquire different viewpoints concerning research in areas in which they have interests, learn about new (unpublished) research, and also learn about various graduate programs.

Superior Knowledge of the Research Process

After supervising student researchers for nearly 40 years, I remain thoroughly convinced that being directly involved in the entire research process, from project development through the convention presentation and publication stages, is the most efficacious way for students to learn about and develop an appreciation for psychological research. The increase in undergraduate research activity in small psychology departments reported by Kierniesky (2005) indicates that this view is becoming more widespread.

One of the courses I taught during this same time period was an experimental psychology/research methods course that required students to conduct an original research project. Hence, I can compare the benefits of this course with the benefits of participating in the lab group under the junior colleague model. Because students in the lab group typically are involved in several complete projects, their knowledge of the research process is superior to that of the students who conduct only one project in order to satisfy the experimental psychology course requirement. (I do not believe that we should stop teaching about research in our classes! Our students do learn a great deal about research in their classes [see Rosell et al., 2005]). The lab group students also have a better appreciation for and love of the research process; however, these differences are likely attributable to the fact that the lab group is self-selected, whereas the experimental psychology course is required of all majors. Students who have conducted several projects and understand the research process are sought after by business and graduate schools. They will require only minimal retraining before they are ready to conduct additional research in new settings.

Superior Knowledge of the Writing of APA Format Papers

Because lab group students are involved in several research projects, they have extensive experience writing papers in accepted APA format. The old adage that "practice makes perfect" is definitely true when it comes to learning the intricacies of APA format!

BENEFITS FOR FACULTY

Students are not the only ones to benefit from student–faculty research collaboration. The faculty also reaps several significant benefits.

Staying Current With the Literature

If you allow students to determine the nature of even a small portion of the research you conduct with students, then you will find yourself doing background reading in a wide range of areas. My experiences indicate that you will be surprised at the wide range of topics your students propose and the diversity of our field. I have found that learning about new and unfamiliar areas is very stimulating; I never cease to be amazed at how interesting and stimulating the field of psychology is. Without question, the new knowledge that you are acquiring via reading the primary literature also will have a positive influence on the classes you teach. In short, this is a win–win situation for everyone.

Networking

When you accompany your research students to conventions where they present a paper and research posters, you also have the opportunity to expand your own professional network by interacting with other faculty from various institutions. Because many of the other faculty will share a common interest, working with students on research, you should be able to cultivate new friendships with ease.

Pride in Seeing Your Students Succeed

From my perspective, pride in seeing your students succeed is the most important benefit or support for faculty. If you have not helped a student bring a research project to fruition and then been in the audience when that student presented the results of his or her project at a convention, you are in for a truly memorable experience—an experience that never loses its excitement. Recently, I experienced a whole new level of pride and excitement when a student of one of my former lab group students presented her first convention paper. I hope that each one of you also will be able to share this experience.

CONCLUDING THOUGHTS

I hope that I have been successful in my efforts to convince you of the benefits and enjoyment, for both students and faculty, of engaging in collaborative research ventures. The benefits and enjoyment are both significant and very real.

I would, however, offer one final caveat: Do not attempt to reproduce exactly my version of the junior colleague model at your institution. What worked for me may not work as well for you. Times are different and different times bring different factors and constraints. Supervising 30+ student researchers who are working simultaneously on five or six projects may not be your cup of tea. Trying to keep up with the burgeoning literature in several unrelated areas might be an insurmountable task. In short, if any of the reflections and suggestions I have made interest you, please adapt them to your own particular situation. Good luck!

This discussion of student–faculty research collaboration recalls the sage words of Professor Charles Brewer, who indicated that "Henry Brooks Adams was right when he said that teachers affect eternity; they never know where their influence ends. However, you must learn to be patient with your students and especially with yourself" (Brewer, 2002, p. 506).

REFERENCES

Brewer, C. L. (2002). Reflections on an academic career: From which side of the looking glass? In S. F. Davis & W. Buskist (Eds.), *The teaching of psychology: Essays in honor of Wilbert J. McKeachie and Charles L. Brewer* (pp. 499–507). Mahwah, NJ: Lawrence Erlbaum Associates.

Dunn, D. S., & Toedter, L. J. (1991). The collaborative honors project in psychology: Enhancing student and faculty development. *Teaching of Psychology, 18,* 178–180.

Evans, R. I., Rintala, D. H., Guthrie, T. J., & Raines, B. E. (1981). Recruiting and training undergraduate psychology research assistants for longitudinal field investigations. *Teaching of Psychology, 8,* 97–100.

Gibson, P. R., Kahn, A. S., & Mathie, V. A. (1996). Undergraduate research groups: Two models. *Teaching of Psychology, 23*, 36–38.

Keith-Spiegel, P., Tabachnick, B. G., & Spiegel, G. B. (1994). When demand exceeds supply: Second-order criteria used by graduate school selection committees. *Teaching of Psychology, 21*, 79–81.

Keith-Spiegel, P., & Wiederman, M. W. (2000). *The complete guide to graduate school admission: Psychology, counseling, and related professions* (2nd ed.) Mahwah, NJ: Lawrence Erlbaum Associates.

Kierniesky, N. C. (1984). Undergraduate research in small psychology departments. *Teaching of Psychology, 11*, 15–18.

Kierniesky, N. C. (2005). Undergraduate research in small psychology departments: Two decades later. *Teaching of Psychology, 32*, 84–90.

Landrum, R. E., & Davis, (2004). *The psychology major: Career options and strategies for success* (2nd ed.). Upper Saddle River, NJ: Prentice-Hall.

Landrum, R. E., & Nelsen, L. R. (2002). The undergraduate research assistant: An analysis of the benefits. *Teaching of Psychology, 29*, 15–19.

McKeachie, W. J. (2002). *Teaching tips: Strategies, research, and theory for college and university teachers* (11th ed.). Boston: Houghton Mifflin.

Rosell, M. C., Beck, D. M., Luther, K. E., Goedert, K. M., Shore, W. J., & Anderson, D. D. (2005). The pedagogical value of experimental participation paired with course content. *Teaching of Psychology, 32*, 95–99.

Slattery, J. M., & Park, C. L. (2002, Spring). Predictors of successful supervision of undergraduate researchers by faculty. *Eye on Psi Chi, 6(3)*, 29–33.

Starke, M. C. (1985). A research practicum: Undergraduates as assistants in psychological research. *Teaching of Psychology, 12*, 158–160.

Chapter **3**

Changing Topics and Trends in Introductory Statistics[1]

Roger E. Kirk
Baylor University

I began teaching introductory statistics over 40 years ago. The course I teach today bears little resemblance to the one I taught then. I now teach material to sophomores that used to be reserved for graduate students, and the number of topics that I cover has increased exponentially. Furthermore, the way I allocate my class time has changed. I now spend much less time on the "how to" aspects of statistics and much more time on the "what it means" aspects. Computers, statistical software, and inexpensive calculators are responsible for many of the changes in my course.

NULL HYPOTHESIS SIGNIFICANCE TESTING: SOME LIMITATION

Before I get into more specific changes, I want to examine an ongoing debate that will have a profound effect on the introductory statistics course. I am referring to the controversy surrounding null hypothesis significance testing. Hypothesis testing as we know it today was developed between 1915 and 1933 by three men: Ronald A. Fisher, Jerzy Neyman, and Egon Pearson. Fisher, who was employed as a statistician at a small agricultural research station 25 miles north of London, was primarily responsible for the new paradigm and for advocating .05 as the standard level of significance (Lehmann, 1993). For over 70 years, null hypothesis significance testing has been the cornerstone of research in psychology. Cohen (1990) observed,

[1] This chapter is based on my keynote address presented at the third annual conference on Best Practices in Teaching Psychology, sponsored by The Society for the Teaching of Psychology, the National Institute for Teaching of Psychology, and the Kennesaw State University Center for Excellence in Teaching and Learning.

Correspondence concerning this article should be addressed to Roger E. Kirk, Baylor University, Waco, TX 76798-7334. Electronic mail can be sent to Roger_Kirk@baylor.edu

> The fact that Fisher's ideas quickly became *the* basis for statistical inference in the behavioral sciences is not surprising—they were very attractive. They offered a deterministic scheme, mechanical and objective, independent of content, and led to clear-cut yes–no decisions. (p. 1307)

In spite of these apparent advantages, null hypothesis significance testing has been surrounded by controversy (Nickerson, 2000). The acrimonious exchanges between Fisher and Neyman that began in 1935 set the pattern for the debate that continues to this day. One of the earliest serious challenges to the logic and usefulness of null hypothesis significance testing appeared in a 1938 article by Joseph Berkson. Since then there has been a crescendo of challenges (Bakan, 1966; Carver 1978, 1993; Cohen, 1990, 1994; Falk & Greenbaum, 1995; Hunter, 1997; Meehl, 1967; Rozeboom, 1960; Schmidt, 1996; Shaver, 1993). The continuing debate eventually led the APA Board of Scientific Affairs to convene a committee called the Task Force on Statistical Inference. Among other things, the committee's charge was "to elucidate some of the controversial issues surrounding applications of statistics including significance testing and its alternatives ..." (Wilkinson & the Task Force on Statistical Inference, 1999, p. 594). The committee's report was published in the *American Psychologist* in 1999 and led to extensive revisions of the statistical sections of the *Publication Manual of the American Psychological Association* (APA, 2001). The following quotation from the Manual captures the essence of the Task Force's recommendations:

> Neither of the two types of probability value [significance level and *p* value] directly reflects the magnitude of an effect or the strength of a relationship. For the reader to fully understand the importance of your findings, it is almost always necessary to include some index of effect size or strength of relationship in your Results section .... The general principle to be followed, however, is to provide the reader not only with information about statistical significance but also with enough information to assess the magnitude of the observed effect or relationship. (pp. 25–26)

Nickerson (2000) provided a detailed summary of the controversy surrounding null hypothesis significance testing. I focus on four frequently mentioned criticisms. Cohen (1994) and others (Berger & Berry, 1988; Carver, 1978; Dawes, Mirels, Gold, & Donahue, 1993; Falk, 1998) criticized the procedure on the grounds that it does not tell researchers what they want to know. To put it another way, null hypothesis significance testing and scientific inference address different questions. In *scientific inference*, what we want to know is the probability that the null hypothesis is true given that we have obtained a set of data, $p(H_0 | D)$. What *null hypothesis significance testing* tells us is the probability of obtaining these data or more extreme data if the null hypothesis is true, $p(D | H_0)$. Researchers incorrectly reason that if the *p* value associated with a test statistic is suitably small, say less than .05, the null hypothesis is probably false. Falk and Greenbaum (1995) referred to this form of deductive reasoning as the "illusion of probabilistic proof by contradiction." Associated with this form of reasoning are two incorrect, widespread beliefs among researchers in psychology: First, the *p* value is the probability that the null hypothesis is true and, second, the complement of the *p* value is the probability that the alternative hypothesis is true. These errors also appear in some introductory statistics textbooks. Nickerson (2000) summarized other common misconceptions regarding null hypothesis significance testing. Here is his list: (a) a small *p* value is indicative of a large treatment effect, (b) the complement of the *p* value is the probability that a significant result will be found in a replication, (c) statistical significance is indicative of practical significance, (d) failure to reject the null hypothesis is equivalent to demonstrating that it is true, and (e) when the probability of observing the data under the null hypothesis is small, the probability of observing the data under the alternative hypothesis must be large.

A second criticism of null hypothesis significance testing is that it is a trivial exercise. In 1991, John Tukey wrote, "the effects of A and B are always different–in some decimal place–for any A and B. Thus asking 'Are the effects different?' is foolish" (p. 100). More recently, Jones and Tukey (2000) reiterated this view,

> For large, finite, treatment populations, a total census is at least conceivable, and we cannot imagine an outcome for which $\mu_A - \mu_B = 0$ when the dependent variable (or any other variable) is measured to an indefinitely large number of decimal places .... The population mean difference may be trivially small but will always be positive or negative. (p. 412)

Tukey (1991) went on to say that a nonrejection of a statistical hypothesis simply means that the researcher is unable to specify the direction of the difference between the conditions. On the other hand, a rejection means that the researcher is pretty sure of the direction of the difference.

The view that null hypotheses are never true except those we construct for Monte Carlo tests of statistical procedures is not new. Frank Yates, a contemporary of Fisher, observed that the null hypothesis, as usually expressed, is "certainly untrue" (Yates, 1964, p. 320). Hence, because all null hypotheses are false, Type I errors cannot occur and statistically significant results are assured if large enough samples are used. Thompson (1998) captured the essence of this view when he wrote, "statistical testing becomes a tautological search for enough participants to achieve statistical significance. If we fail to reject, it is only because we've been too lazy to drag in enough participants" (p. 799). Cohen (1962, 1969) noted that it is ironic that a ritualistic adherence to null hypothesis significance testing has led researchers to focus on controlling the Type I error that cannot occur because all null hypotheses are false while allowing the Type II error that can occur to exceed acceptable levels, often as high as .50 to .80.

A third criticism of null hypothesis significance testing is that by adopting a fixed level of significance, a researcher turns a continuum of uncertainty into a dichotomous reject/do-not-reject decision (Frick, 1996;

Grant, 1962; Rossi, 1997; Wickens, 1998). A *p* value only slightly larger than the level of significance is treated the same as a much larger *p* value. Some researchers attempt to blur the reject/do-not-reject dichotomy with phrases such as "the results approached significance" or "the results were marginally significant." However, studies of the way applied researchers interpret *p* values find a "cliff effect" at .05 in which the reported confidence in research findings drops perceptibly when *p* becomes larger than .05 (Beauchamp & May, 1964; Rosenthal & Gaito, 1963, 1964). The adoption of .05 as the dividing point between significance and nonsignificance is quite arbitrary. Rosnow and Rosenthal's (1989) comment is relevant: "Surely, God loves the .06 nearly as much as the .05" (p. 1277).

A fourth criticism of null hypothesis significance testing is that it does not address the question of whether results are important, valuable, or useful—that is, their practical significance. Practical significance is concerned with the usefulness of results; statistical significance, the focus of null hypothesis significance tests, is concerned with whether results are due to chance or sampling variability (Kirk, 1996).

In response to these criticisms, researchers have suggested ways to improve null hypothesis testing procedures. For example, Serlin (1993) and others suggested that instead of testing a point null hypothesis that is always false, researchers should test a range null hypothesis (Serlin & Lapsley, 1985, 1993; Yelton & Sechrest, 1986). Another suggestion is the three-outcome test. In this modification, the alternative hypothesis, say $\mu_1 \neq \mu_2$, is replaced with two directional alternatives (Bohrer, 1979; Harris, 1994, 1997; Jones & Tukey, 2000; Kaiser, 1960). Yet another suggestion is a modified split-tail test (Braver, 1975; Nosanchuk, 1978). These suggestions and others have found little acceptance among researchers and have had no impact on the introductory statistics course.

It is clear that null hypothesis significance testing is open to criticisms. But, unfortunately, most authors of introductory statistics books do not mention these and other criticisms. In a convenience sample of eight introductory statistics textbooks published in 2003 or later, I found only two that mentioned one or more criticisms of null hypothesis significance testing. To not discuss the limitations of null hypothesis significance tests is to do a disservice to our students. Students need to understand what a significance test does and does not tell a researcher. Clearly, textbook authors need to do better in this area. I expect that in the near future, authors will provide a more balanced presentation of null hypothesis significance testing.

SUPPLEMENTING NULL HYPOTHESIS SIGNIFICANCE TESTS

Reporting Effect Sizes

Earlier, I quoted one of the recommendations that appears in the 2001 APA *Publication Manual*, "provide the reader not only with information about statistical significance but also with enough information to assess the magnitude of the observed effect or relationship" (APA, 2001, p. 26). I was surprised to discover that 40 years ago, Frank Yates (1964) anticipated this *Publication Manual* recommendation. In discussing significance testing, he said that one of its weakness is a "failure to recognize that in many types of experimental work estimates of the treatment effects, together with estimates of the errors to which they are subject, are the quantities of primary interest" (p. 320). Both the *Publication Manual* and Yates recognized the need to supplement the information provided by null hypothesis significance tests.

Researchers can use a variety of statistics to supplement significance tests and assess the magnitude of observed effects. Many of the statistics fall into one of two categories: measures of effect size—typically, standardized mean differences—and measures of strength of association. In addition, there is a large group of statistics that do not fit into either category. Among researchers who work in substantive areas, interest in measures of effect magnitude has increased over the last 30 years. I published an article in 1996 in which I listed over 40 measures of effect magnitude used in psychology and education journals (Kirk, 1996). I recently updated the list; it contains 72 measures (Kirk, 2005).

How do substantive researchers use measures of effect magnitude? I have identified four major uses: (a) estimating the sample size required to achieve an acceptable power, (b) integrating the results of a number of studies in a meta-analysis, (c) supplementing the information provided by null hypothesis significance tests and confidence intervals, and (d) determining whether research results are practically significant (Kirk, 2003).

The first measure of effect size that was explicitly labeled as such appeared in the social and behavioral sciences literature in 1969. The measure, δ, was developed by Jack Cohen and expresses the size of a population contrast of means in units of the population standard deviation as shown in the following equation:

$$\delta = \frac{\psi}{\sigma} = \frac{\mu_E - \mu_C}{\sigma},$$

where ψ denotes a contrast between the population mean of the experimental group, μ_E, and the control group, μ_C, and σ denotes the common population standard deviation. Cohen noted that the size of a contrast is influenced by the scale of measurement of the means. He divided the contrast by the population standard deviation to rescale the contrast in units of the amount of error variability in the data. Rescaling is useful when the measurement units are arbitrary or have no inherent meaning. Rescaling also can be useful in performing power and sample size computations and in comparing effect sizes across research literatures involving diverse, idiosyncratic measurement scales. However, rescaling serves no purpose when a variable is always measured on a standard scale.

For nonstandard measurement scales, Cohen's contribution is significant because he provided guidelines for interpreting the magnitude of δ.

He said that $\delta = 0.2$ is a small effect, $\delta = 0.5$ is a medium effect, and $\delta = .8$ is a large effect. According to Cohen (1992), a medium effect of .5 is visible to the naked eye of a careful observer. A small effect of .2 is noticeably smaller than medium but not so small as to be trivial. Only an expert would be able to detect a small effect. A large effect of .8 is the same distance above medium as small is below it. A large effect would be obvious to anyone. Several surveys have found that .5 approximates the average size of observed effects in a number of fields (Cooper & Findley, 1982; Haase, Waechter, & Solomon, 1982; Sedlmeier & Gigerenzer, 1989).

By assigning the labels small, medium, and large to the numbers .2, .5, and .8, respectively, Cohen provided researchers with guidelines for interpreting the size of treatment effects. His effect size measure is a valuable supplement to the information provided by a *p* value. A *p* value of .0001 loses its luster if the effect size turns out to be less than .2. Effect sizes also are useful for comparing and integrating the results of different studies in a meta-analysis.

The parameters of Cohen's δ are rarely known. The sample means of the experimental and control groups are used to estimate the population means. An estimator of σ can be obtained in a number of ways. Under the assumption that the population standard deviations of the experimental and control groups are equal, Cohen pooled the two sample variances to obtain an estimator of δ as follows:

$$d = \frac{\overline{Y}_E - \overline{Y}_C}{\hat{\sigma}_{Pooled}},$$

where

$$\hat{\sigma}_{Pooled} = \sqrt{\frac{(n_E - 1)\hat{\sigma}_E^2 + (n_C - 1)\hat{\sigma}_C^2}{(n_E - 1) + (n_C - 1)}}.$$

Hedges (1981) noted that the population standard deviations of $p \geq 2$ samples are often homogeneous, in which case the best estimator of σ is obtained by pooling all of the variances. His estimator of δ is

$$g = \frac{\overline{Y}_E - \overline{Y}_C}{\hat{\sigma}_{Pooled}},$$

where

$$\hat{\sigma}_{Pooled} = \sqrt{\frac{(n_1 - 1)\hat{\sigma}_1^2 + (n_2 - 1)\hat{\sigma}_2^2 + \ldots + (n_p - 1)\hat{\sigma}_p^2}{(n_1 - 1) + (n_2 - 1) + \ldots + (n_p - 1)}}.$$

This pooled estimator is identical to the square root of the within-groups mean square in a completely randomized analysis of variance. Hedges observed that γ is biased. He recommended correcting g for bias as follows,

$$g_c = J(N-2)g,$$

where J(N–2) is the bias correction factor described in Hedges and Olkin (1985) and $N = n_1 + n_2 + \ldots + n_p$. The correction factor is approximately

$$J(N-2) \cong \left(1 - \frac{3}{4N-9}\right).$$

Hedges also described an approximate confidence interval for δ:

$$g_c - z_{\alpha/2}\,\hat{\sigma}(g_c) \leq \delta \leq g_c + z_{\alpha/2}\,\hat{\sigma}(g_c),$$

where $z_{\alpha/2}$ denotes the two-tailed critical value that cuts off the upper $\alpha/2$ region of the standard normal distribution and

$$\hat{\sigma}(g_c) = \sqrt{\frac{n_1 + n_2}{n_1 n_2} + \frac{g_c^2}{2(n_1 + n_2)}}.$$

Cumming and Finch (2001) described procedures for obtaining exact confidence intervals using noncentral sampling distributions.

Effect size is an example of one of the changing topics in statistics courses. Examine any recently published introductory statistics textbook and you will find a discussion of Cohen's effect size. This was not true 15 years ago. In a convenience sample of eight texts published between 1989 and 1991, I found only two that provided a discussion of effect size.

Cohen and Hedges' estimators of δ have a number of features that contribute to their popularity: (a) they are easy to understand and have a consistent interpretation across different research studies, (b) the sampling distributions of the estimators are well understood, and (c) the estimators can be readily computed from published articles that report t statistics and F statistics with one degree of freedom. The latter feature is particularly attractive to researchers who do meta-analyses. However, there are problems. The values of d and g are affected by the amount of variability in the samples. Hence, estimates of δ from different experiments are not comparable unless the experiments have similar standard deviations. Also, the correct way to conceptualize and compute the denominator of δ can be problematic when the treatment is a classification or organismic variable (Gillett, 2003; Grissom & Kim, 2001; Olejnik & Algina, 2000). For experiments with a manipulated treatment and random assignment of participants to the treatment levels, the computation of an effect size such as g_c is relatively straightforward. The denominator of g_c is the square root of the within-groups mean square. This mean square provides an estimate of σ that reflects the variability of observations for the full range of the manipulated treatment. However, when the treatment is an organismic variable such as gender, say boys and girls, the square root of the within-groups mean square may not reflect the variability for the full range of the treatment because it is a pooled measure of the variation of boys alone and the variation of girls alone. If there is a gender effect, the within-groups mean square reflects the variation for a partial range of the gender variable. The variation for the full range of the gender variable is given by the total mean square and will be larger than the within-groups mean square. Effect sizes should be comparable across different kinds of treatments and experimental designs. In the gender experiment, use of the square root of the total mean square to estimate σ gives an effect size that is comparable to those for treatments that are manipulated.

The problem of estimating σ is exacerbated when the experiment has several treatments, repeated measures, and covariates. Gillett (2003) and Olejnik and Algina (2000) provided guidelines for computing effect sizes for such designs. And there are other problems. The estimators assume normality and a common standard deviation. Furthermore, the value of the estimators is greatly affected by heavy-tailed distributions and heterogeneous standard deviations (Wilcox, 1996). These problems have led some researchers to attempt to improve the estimation of δ; other researchers have focused on alternative ways of conceptualizing effect magnitude (Hays, 1963; Hedges & Olkin, 1985; Kendall, Marss-Garcia, Nath, & Sheldrick, 1999; Kraemer, 1983; Lax, 1985; Olejnik & Algina, 2000; Wilcox, 1996, 1997). I briefly describe two of these measures next.

Reporting Strength of Association

The idea of supplementing the null hypothesis significance test with a measure of strength of association as recommended by the APA *Publication Manual* is not new. In 1925 Fisher made the same recommendation. He proposed that researchers supplement the significance test in analysis of variance with the *correlation ratio*, which is a measure of effect magnitude.

Two popular measures of strength of association are omega squared, ω^2, for a fixed-effects treatment and the intraclass correlation, ρ_I, for a random-effects treatment. For a completely randomized analysis of variance design, ω^2 and ρ_I are defined as

$$\frac{\sigma^2_{Treat}}{\sigma^2_{Treat} + \sigma^2_{Error}},$$

where σ^2_{Treat} and σ^2_{Error} denote, respectively, the population variances for treatment and error. According to Hays (1963), who introduced omega squared, ω^2 and ρ_I indicate the proportion of the population variance in the dependent variable that is accounted for by specifying the treatment-level classification and thus are identical in general meaning. Formulas for estimating ω^2 and ρ_I differ because the expected values of the treatment mean squares for the fixed- and random-effects models differ. Estimators of ω^2 and ρ_I are

$$\hat{\omega}^2 = \frac{SS_{Treat} - (p-1)MS_{Error}}{SS_{Total} + MS_{Error}},$$

$$\hat{\rho}_I = \frac{MS_{Treat} - MS_{Error}}{MS_{Treat} + (n-1)MS_{Error}},$$

where *SS* and *MS* denote, respectively, sum of squares and mean squares, *p* denotes the number of treatment levels, and *n* denotes the number of observations in each treatment level. Both omega squared and the intraclass correlation are biased estimators. However, Carroll and Nordholm (1975) showed that the degree of bias in omega squared computed from sample data is slight. Earlier, I noted that the usefulness of Cohen's δ was enhanced when he suggested guidelines for its interpretation. Cohen (1969) also suggested guidelines for interpreting squared correlation coefficients: $\omega^2 = .010$ is a small association, $\omega^2 = .059$ is a medium association, and $\omega^2 = .138$ is a large association.

Omega squared and the intraclass correlation, like measures of effect size, are not without their detractors. One criticism voiced by O'Grady (1982) is that omega squared and the intraclass correlation may underestimate the true proportion of explained variance. If, as is generally the case, the dependent variable is not perfectly reliable, measurement error will reduce the proportion of variance that can be explained. O'Grady also criticized measures of strength of association on the grounds that their value is affected by the choice and number of treatment levels. In general, the greater the diversity and number of treatment levels, the larger is the strength of association. Joel Levin (1967) observed that omega squared is not very informative when an experiment contains more than two treatment levels. A large value of omega squared simply indicates that the dependent variable for at least one treatment level is substantially different from that of the other levels. As is true for all omnibus measures, omega squared and the intraclass correlation do not pinpoint which treatment level(s) is responsible for a large value.

One way to address the last criticism is to compute omega squared and the intraclass correlation for two-mean contrasts as is done with Cohen's effect size. This solution is in keeping with the preference of many quantitative psychologists to ask focused one-degree-of freedom questions of their data (Judd, McClelland, & Culhane 1995; Rosnow, Rosenthal, & Rubin, 2000) and the recommendation of the *Publication Manual of the American Psychological Association* (2001). The *Manual* states, "As a general rule, multiple degree-of-freedom effect indicators tend to be less useful than effect indicators that decompose multiple degree-of-freedom tests into meaningful one degree-of-freedom effects—particularly when these are the results that inform the discussion" (p. 26). The formulas for omega squared and the intraclass correlation can be modified to give the proportion of variance in the dependent variable that is accounted for by the ith contrast, $\hat{\psi}_i$. The formulas are as follows

$$\hat{\omega}^2_{Y|\psi_i} = \frac{SS\hat{\psi}_i - MS_{Error}}{SS_{Treat} + MS_{Error}},$$

$$\hat{\rho}_{I\ Y|\psi_i} = \frac{SS\hat{\psi}_i - MS_{Error}}{SS\hat{\psi}_i + (n-1)MS_{Error}}$$

where $SS\hat{\psi}_i = \hat{\psi}_i^2 \ / \Sigma_{j=1}^{p} c_j^2 \ / n_j$ and the c_j's are coefficients that define a contrast (Kirk 1995). These two measures answer focused one-degree-of-freedom questions as opposed to omnibus questions about one's data. Neither of these measures appeared in a convenience sample of eight introductory statistics texts published between 1989 and 1991. In a convenience sample of seven texts published in 2003 or later, I found two that discussed the *omnibus* omega squared statistic. However, none of the texts described the one-degree-of freedom measures. Will either of these statistics find their way into the introductory statistics course? I do not know. I do know that simply covering Cohen's d is no longer sufficient. Because of APA's recommendations, I expect that the coverage of alternative measures will grow. Clearly, we still have a long way to go in implementing the recommendations of the 1999 APA Task Force report.

I have described two categories of effect magnitude measures: *measures of effect size* and *strength of association*. Researchers differ in their preferences for the measures. Fortunately, it is a simple matter to convert from one measure to another. To make conversions among d (g), r_{pb}, $\hat{\omega}^2_{Y|\psi_i}$ and $\hat{\rho}_{I\ Y|\psi_i}$ or to convert Student's t statistic into any of the four measures, see Kirk (2005) for conversion formulas. To fully appreciate the importance of research results, students need to be able to convert Student's t into a measure of effect magnitude. Only one of the seven texts published in 2003 or later provided a formula for converting a t statistic into Cohen's d. In the next few years, I expect that more texts will provide conversion formulas.

ADVANTAGES OF CONFIDENCE INTERVALS

Another recommendation in the APA *Publication Manual* (2001) is just beginning to affect the introductory statistics course. I am referring to the recommendation regarding confidence intervals. The *Manual* states:

> The reporting of confidence intervals (for estimates of parameters, for functions of parameters such as differences in means, and for effect sizes) can be an extremely effective way of reporting results. Because confidence intervals combine information on location and precision and can often be directly used to infer significance levels, they are, in general, the best reporting strategy. The use of confidence intervals is therefore strongly recommended. (APA, 2001, p. 22)

Confidence intervals and null hypothesis significance tests are two complementary approaches to classical statistical inference. An important advantage of a confidence interval is that it requires the same assumptions and data as a null hypothesis significance test, but the interval provides much more information. As John Tukey pointed out in 1991, the rejection of a null hypothesis is not very informative. A researcher knows in advance that the hypothesis is false. However, a descriptive statistic and confidence interval provide an estimate of the population parameter and a range of values—the error variation—qualifying that estimate. A 95% confidence interval for, say $\mu_1 - \mu_2$, contains all the values for which the null hypothesis would not be rejected at the .05 level of significance. Values outside the confidence interval would be rejected. Furthermore, a descriptive statistic and confidence interval use the same unit of measurement as the data. This facilitates the interpretation of results and makes trivial effects more obvious (Kirk, 2001).

Confidence intervals and measures of effect magnitude are especially useful in assessing the practical significance of results. Currently, what we see in our journals is the use of a reject–nonreject decision strategy that does not tell researchers what they want to know and a preoccupation with *p* values that are several steps removed from examining the data. I expect this situation will change in the next few years. What is surprising is that authors of introductory statistics textbooks have largely ignored the 1999 Task Force report and the 2001 APA *Publication Manual* recommendations regarding confidence intervals. This is apparent when we compare the coverage of confidence intervals in the convenience sample of eight introductory statistics books published between 1989 and 1991 with the coverage of seven books published between 2003 and 2005. As shown in Table 3.1, all eight introductory statistics books published prior to the 1999 APA Task Force report included a discussion of confidence intervals. However, the coverage varied greatly from one text to the next. Four of the texts discussed only a confidence interval for a population mean. The convenience sample of seven textbooks published four or more years after the 1999 Task Force report appear in the bottom portion of Table 3.1. These books provide a bit more coverage of confidence intervals than the books published prior to 1999. However, considering the strong recommendations in the APA Task Force report and the 2001 APA *Publication Manual*, I would have expected to see a much greater coverage. Most of the recently published books limit their coverage to confidence intervals for a population mean and contrasts for two

TABLE 3.1
Coverage of Confidence Intervals in Fifteen Introductory Statistics Textbooks

	μ	*Mdn*	p	ρ	*Regression* $\hat{Y}$	*Independent* $\mu_1-\mu_2$	*Dependent* $\mu_1-\mu_2$	*Independent* p_1-p_2	*Dependent* p_1-p_2
					Textbooks Published Between 1989 and 1991				
Text 1 (1991)	Yes								
Text 2 (1990)	Yes								
Text 3 (1989)	Yes								
Text 4 (1990)	Yes								
Text 5 (1991)	Yes	Yes				Yes			
Text 6 (1989)	Yes					Yes	Yes		
Text 7 (1989)	Yes	Yes	Yes						
Text 8 (1990)	Yes		Yes			Yes	Yes	Yes	Yes
					Textbooks Published Between 2003 and 2005				
Text 1 (2005)	Yes								
Text 2 (2004)	Yes					Yes	Yes		
Text 3 (2003)	Yes					Yes	Yes		
Text 4 (2004)	Yes		Yes			Yes	Yes		
Text 5 (2004)	Yes			Yes		Yes	Yes		
Text 6 (2003)	Yes				Yes	Yes	Yes		
Text 7 (2003)	Yes		Yes	Yes	Yes	Yes	Yes	Yes	Yes

means. What is particularly disappointing is that most of these textbooks fail to point out the advantages of confidence intervals and only two of the books point out one or more limitations of null hypothesis significance tests. With one exception, the books follow the common practice of describing hypothesis tests before confidence intervals.

THE EXPANDING CONTENT OF INTRODUCTORY STATISTICS BOOKS

Earlier I mentioned that the number of topics that I cover in my introductory statistics course has increased. I see a similar increase in the number and level of topics that are covered in introductory statistics books published between 2003 and 2005. Fifteen years ago, multiple regression was mentioned in only one of the eight statistics books that I examined. All of the recently published books mention multiple regression. One book devoted an entire chapter to the topic.

The inclusion of more advanced statistical topics in introductory statistics books parallels the increasing availability of statistical packages. Table 3.2 illustrates the trend. The inclusion of some advanced topics is surprising. For example, two of the books describe multivariate analysis of variance and one book even provided a description of structural equation modeling. I think that this trend reflects a changing orientation in our introductory statistics courses. Years ago, we were satisfied to simply introduce students to basic statistics. However, our message to undergraduates over the years has been that they need to take a statistics course in order to read psychology journals. But our journals are using more and more sophisticated techniques. As a result, authors apparently feel a need to describe these advanced techniques in their books.

It seems that advances in technology inevitably fuel changes in the introductory statistics course. I expect that the Internet and World Wide Web also will affect the way we teach the course. Currently, instructors use these resources mainly to provide students with (a) additional practice problems and review exercises, (b) interactive demonstrations, (c) computing tools, and (d) access to large data sets. It is hard for me to imagine the diverse ways in which online information directories and search engines will be used in the future. It is clear, however, that students will have easy access to an amazing amount of statistical information.

It also is clear that advances in technology—the Web, calculators, statistical packages, multimedia classrooms, and so on—are affecting the way we allocate our class time. My colleagues around the country tell me that they spend less time teaching students how to compute things and more time explaining what things mean. I have experienced the same changes. And with presentation software such as Power Point™, they spend less time writing on blackboards. A number of my colleagues also have told me that they have stopped teaching raw score formulas. Instead they focus on definitional or deviation formulas. This is one way to gain more time to cover the breath of new topics that are appearing in textbooks.

TABLE 3.2

Coverage of Advanced Topics in Eight Introductory Statistics Textbooks Published Between 2003 and 2005

	Multiple Regression	*Logistic Regression*	*Partial or Part Correlation*	*ANCOVA*	*Multivariate ANOVA*	*Structural Equation Modeling*
Text 1 (2005)	Yes		Yes	Yes	Yes	Yes
Text 2 (2004)	Yes	Yes		Yes	Yes	
Text 3 (2003)	Yes					
Text 4 (2004)	Yes					
Text 5 (2004)	Yes	Yes		Yes		
Text 6 (2003)	Yes		Yes			
Text 7 (2003)	Yes					

THE IMPACT OF CHANGING DEMOGRAPHICS

In the 40-plus years that I have been teaching introductory statistics, I have seen a major shift in the demographics of my students. When I began teaching at Baylor University, 85% of the psychology majors were men. Now the numbers are reversed; 85% of our majors are women. In the past, our doctoral students were about equally divided between the experimental and clinical areas. All of us are aware of the change in this statistic. According to the most recent data from the American Psychological Association (2005), 47% of doctorates are in clinical areas. Experimental psychology accounts for only 2%. These figures have implications for the introductory statistics course. We have to recognize that most of the students in our classes will be consumers of statistics rather than primary users of statistics. Today, my students tend to be less research oriented, more people oriented, and have poorer math skills than my students 40 years ago. Furthermore, they dread taking my course and fail to see its relevance.

I documented these and other attitudes about my statistics course years ago when I began administering a sentence completion attitude survey on the first and last day of the semester. The survey measures such things as attitudes toward taking the course, the place of statistics in a student's career goals, and so on. I tell the students that there are no right or wrong answers and to complete each sentence with their first thought. What I discovered from the surveys is that almost all of the students dread taking my course. A random sample of 15 responses from my fall semester class to the item "When I realized that I had to take this course, I ..." is as follows:

was scared; panicked, I have heard horror stories about this course; knew I'd have to take it twice; didn't understand what it had to do with psychology; asked my father for help since he is a psychologist; didn't have any idea what to expect, everyone has told me that it's a killer; was anxious; nervous but interested; thought it would be a definite challenge for me; knew it would be hard; was apprehensive about the grading scale; knew I'd be facing a challenge; dreaded it, because of the rumors I heard about the class; wasn't too thrilled; thought that if I wanted to be a psychologist I would have to take it and suffer through it; just accepted it since I had not heard anything about it.

Each semester as I analyze the survey results, I feel a twinge of envy for my colleagues who teach abnormal psychology and other popular courses.

The widely held perception among my students is that statistics is not psychology and hence is not relevant to their career goals. Once I realized just how widespread this perception was, I modified my teaching approach. Because of the changing demographics, I now spend much more time showing how statistics is relevant for psychologist-practitioners and how research using statistics has changed psychological practices. I do this in a number of ways. For example, I introduce each new statistic with a real-life example of the way the statistic is used in psychology. It is not easy to convince students that statistics is relevant to psychology. Another technique I use is to project results and discussion sections of interesting journal articles on two large screens in the multimedia classroom where I teach. I moderate a discussion of the articles. I encourage students to decide whether the author's conclusions are supported by the data and results. The discussions have two important benefits: Students see how the statistics they are learning are used to advance psychological knowledge and they get some practice in critical thinking.

SUMMARY OF TRENDS AND SOME PREDICTIONS

In closing, I want to summarize some of the trends that I see in the introductory statistics course and make some predictions.

1. The range of topics in the introductory course will continue to expand.
2. A related trend, providing simplified descriptions of topics that are normally covered in advanced courses, will continue.
3. Instructors will gain time to cover a wider range of topics by utilizing technologies such computers, multimedia classrooms, the Internet, and Web-based materials.
4. Instructors will focus more on definitional formulas for statistics and the meaning of the statistics and devote less time to raw score formulas and the how-to-compute-it aspects of statistics.
5. In order to motivate students, instructors will make greater use of lifelike data sets that simulate the results of real experiments.
6. Introductory statistics textbooks will begin to point out limitations of null hypothesis significance testing.
7. The coverage of various measures of effect magnitude will increase dramatically.
8. The coverage of confidence intervals will increase, but instructors will continue to emphasize null hypothesis significance testing.
9. The aging of the membership of APA's Division 5 (Evaluation, Measurement, and Statistics) will continue. As a result, more and more students will receive their statistics instruction from professors whose specialty is not statistics. This change will place a greater burden on textbook authors to present clearly the methodological foundations of our discipline.
10. The challenge of teaching the introductory statistics course will continue to require great patience, compassion, dedication, and creativity.

REFERENCES

American Psychological Association. (2001). *Publication manual of the American Psychological Association* (5th ed.). Washington, DC: Author.

American Psychological Association. (2005). *Graduate study in psychology*. Washington, DC: Author.
Bakan, D. (1966). The test of significance in psychological research. *Psychological Bulletin, 66*, 423–437.
Beauchamp, K. L., & May, R. B. (1964). Replication report: Interpretation of levels of significance by psychological researchers. *Psychological Reports, 14*, 272.
Berger, J. O., & Berry, D. A. (1988). Statistical analysis and the illusion of objectivity. *American Scientist, 76*, 159–165.
Berkson, J. (1938). Some difficulties of interpretation encountered in the application of the chi-square test. *Journal of the American Statistical Association, 33*, 526–542.
Bohrer, R. (1979). Multiple three-decision rules for parametric signs. *Journal of the American Statistical Association, 74*, 432–437.
Braver, S. L. (1975). On splitting the tails unequally: A new perspective on one-versus two-tailed tests. *Educational and Psychological Measurement, 35*, 283–301.
Carroll, R. M., & Nordholm, L. A. (1975). Sampling characteristics of Kelley's ε^2 and Hays' ω^2. *Educational and Psychological Measurement, 35*, 541–554.
Carver, R. P. (1978). The case against statistical significance testing. *Harvard Educational Review, 48*, 378–399.
Carver, R. P. (1993). The case against statistical significance testing, revisited. *Journal of Experimental Education, 61*, 287–292.
Cohen, J. (1962). The statistical power of abnormal-social psychological research: A review. *Journal of Abnormal and Social Psychology, 65*, 145–153.
Cohen, J. (1969). *Statistical power analysis for the behavioral sciences*. New York: Academic Press.
Cohen, J. (1990). Things I have learned (so far). *American Psychologist, 45*, 1304–1312.
Cohen, J. (1992). A power primer. *Psychological Bulletin, 112*, 115–159.
Cohen, J. (1994). The earth is round ($p < .05$). *American Psychologist, 49*, 997–1003.
Cooper, H., & Findley, M. (1982). Expected effect sizes: Estimates for statistical power analysis in social psychology. *Personality and Social Psychology Bulletin, 8*, 168–173.
Cumming, G., & Finch, S. (2001). A primer on the understanding, use, and calculation of confidence intervals that are based on central and noncentral distributions. *Educational and Psychological Measurement, 61*, 532–574.
Dawes, R. M., Mirels, H. L., Gold, E., & Donahue, E. (1993). Equating inverse probabilities in implicit personality judgments. *Psychological Science, 4*, 396–400.
Falk, R. (1998). Replication—A step in the right direction. *Theory and Psychology, 8*, 313–321.
Falk, R., & Greenbaum, C. W. (1995). Significance tests die hard: The amazing persistence of a probabilistic misconception. *Theory and Psychology, 5*, 75–98.
Fisher, R. A. (1925). *Statistical methods for research workers*. Edinburgh: Oliver & Boyd.
Frick, R. W. (1996). The appropriate use of null hypothesis testing. *Psychological Methods, 1*, 379–390.
Gillett, R. (2003). The comparability of meta-analytic effect-size estimators from factorial designs. *Psychological Methods, 8*, 419–433.

Grant, D. A. (1962). Testing the null hypothesis and the strategy and tactics of investigating theoretical models. *Psychological Review, 69*, 54–61.

Grissom, R. J., & Kim, J. J. (2001). Review of assumptions and problems in the appropriate conceptualization of effect size. *Psychological Methods, 6*, 135–146.

Haase, R. F., Waechter, D. M., & Solomon, G. S. (1982). How significant is a significant difference? Average effect size of research in counseling psychology. *Journal of Counseling Psychology, 29*, 58–65.

Harris, R. J. (1994). *ANOVA: An analysis of variance primer.* Itasca, IL: F. E. Peacock.

Harris, R. J. (1997). Reforming significance testing via three-valued logic. In L. L. Harlow, S. A. Mulaik, & J. H. Steiger (Eds.), *What if there were no significance tests?* (pp. 145–174). Hillsdale, NJ: Lawrence Erlbaum Associates.

Hays, W. L. (1963). *Statistics for psychologists.* New York: Holt, Rinehart & Winston.

Hedges, L. V. (1981). Distributional theory for Glass's estimator of effect size and related estimators. *Journal of Educational Statistics, 6*, 107–128.

Hedges, L. V., & Olkin, I. (1985). *Statistical methods for meta-analysis.* Orlando, FL: Academic Press.

Hunter, J. E. (1997). Needed: A ban on the significance test. *Psychological Science, 8*, 3–7.

Jones, L. V., & Tukey, J. W. (2000). A sensible formulation of the significance test. *Psychological Methods, 5*, 411–414.

Judd, C. M., McClelland, G. H., & Culhane, S. E. (1995). Data analysis: Continuing issues in the everyday analysis of psychological data. *Annual Reviews of Psychology, 46*, 433–465.

Kaiser, H. F. (1960). Directional statistical decisions. *Psychological Review, 67*, 160–167.

Kendall, P. C., Marss-Garcia, A., Nath, S. R., & Sheldrick, R. C. (1999). Normative comparisons for the evaluation of clinical significance. *Journal of Consulting and Clinical Psychology, 67*, 285–299.

Kirk, R. E. (1995). *Experimental design: Procedures for the behavioral sciences* (3rd ed.). Monterey, CA: Brooks/Cole.

Kirk, R. E. (1996). Practical significance: A concept whose time has come. *Educational and Psychological Measurement, 56*, 746–759.

Kirk, R. E. (2001). Promoting good statistical practices: Some suggestions. *Educational and Psychological Measurement, 61*, 213–218.

Kirk, R. E. (2003). The importance of effect magnitude. In S. F. Davis (Ed.), *Handbook of research methods in experimental psychology* (pp. 83–105). Oxford, England: Blackwell.

Kirk, R. E. (2005). Effect size measures. In B. Everitt & D. Howell (Eds.), *Encyclopedia of statistics in behavioral science* (Vol. 2, pp. 532–542). New York: Wiley.

Kraemer, H. C. (1983). Theory of estimation and testing of effect sizes: Use in meta-analysis. *Journal of Educational Statistics, 8*, 93–101.

Lax, D. A. (1985). Robust estimators of scale: Finite sample performance in long-tailed symmetric distributions. *Journal of the American Statistical Association, 80*, 736–741.

Lehmann, E. L. (1993). The Fisher, Neyman-Pearson theories of testing hypotheses: One theory or two. *Journal of the American Statistical Association, 88*, 1242–1248.

Levin, J. R. (1967). Misinterpreting the significance of "explained variation." *American Psychologist, 22*, 675–676.

Meehl, P. E. (1967). Theory testing in psychology and physics: A methodological paradox. *Philosophy of Science, 34*, 103–115.
Nickerson, R. S. (2000). Null hypothesis significance testing: A review of an old and continuing controversy. *Psychological Methods, 5*, 241–301.
Nosanchuk, T. A. (1978). Serendipity tails: A note on two-tailed hypothesis tests with asymmetric regions of rejection. *Acta Sociologica, 21*, 249–253.
O'Grady, K. E. (1982). Measures of explained variation: Cautions and limitations. *Psychological Bulletin, 92*, 766–777.
Olejnik, S., & Algina, J. (2000). Measures of effect size for comparative studies: Applications, interpretations, and limitations. *Contemporary Educational Psychology, 25*, 241–286.
Rosenthal, R., & Gaito, J. (1963). The interpretation of levels of significance by psychological researchers. *Journal of Psychology, 55*, 33–38.
Rosenthal, R., & Gaito, J. (1964). Further evidence for the cliff effect in interpretation of levels of significance. *Psychological Reports, 15*, 570.
Rosnow, R. L., & Rosenthal, R. (1989). Statistical procedures and the justification of knowledge in psychological science. *American Psychologist, 44*, 1276–1284.
Rosnow, R. L., Rosenthal, R., & Rubin, D. B. (2000). Contrasts and correlations in effect-size estimation. *Psychological Science, 11*, 446–453.
Rossi, J. S. (1997). A case study in the failure of psychology as cumulative science: The spontaneous recovery of verbal learning. In L. L. Harlow, S. A. Mulaik, & J. H. Steiger (Eds.), *What if there were no significance tests?* (pp. 175–197). Hillsdale, NJ: Lawrence Erlbaum Associates.
Rozeboom, W. W. (1960). The fallacy of the null hypothesis significance test. *Psychological Bulletin, 57*, 416–428.
Schmidt, F. L. (1996). Statistical significance testing and cumulative knowledge in psychology: Implications for the training of researchers. *Psychological Methods, 1*, 115–129.
Sedlmeier, P., & Gigerenzer, G. (1989). Do studies of statistical power have an effect on the power of studies? *Psychological Bulletin, 105*, 309–316.
Serlin, R. C. (1993). Confidence intervals and the scientific method: A case for Holm on the range. *Journal of Experimental Education, 61*, 350–360.
Serlin, R. C., & Lapsley, D. K. (1985). Rationality in psychological research: The good-enough principle. *American Psychologists, 40*, 73–83.
Serlin, R. C., & Lapsley, D. K. (1993). Rational appraisal of psychological research and the good-enough principle. In G. Keren & C. Lewis (Eds.), *A handbook for data analysis in the behavioral sciences: Methodological issues* (pp. 199–228). Hillsdale, NJ: Lawrence Erlbaum Associates.
Shaver, J. P. (1993). What statistical significance testing is, and what it is not. *Journal of Experimental Education, 61*, 293–316.
Thompson, B. (1998). In praise of brilliance: Where the praise really belongs. *American Psychologist*, 53, 799–800.
Tukey, J. W. (1991). The philosophy of multiple comparisons. *Statistical Science*, 6, 100–116.
Wickens, C. D. (1998). Commonsense statistics. *Ergonomics in Design*, 6(4), 18–22.
Wilcox, R. R. (1996). *Statistics for the social sciences*. San Diego, CA: Academic Press.
Wilcox, R. R. (1997). *Introduction to robust estimation and hypothesis testing*. San Diego, CA: Academic Press.

Wilkinson, L., & the Task Force on Statistical Inference. (1999). Statistical methods in psychology journals: Guidelines and explanations. *American Psychologists, 54*, 594–604.

Yates, F. (1964). Sir Ronald Fisher and the design of experiments. *Biometrics, 20*, 307–321.

Yelton, W. H., & Sechrest, L. (1986). Use and misuse of no-difference findings in eliminating threats to validity. *Evaluation Review, 10*, 836–852.

Chapter 4

Creating Syllabi for Statistics and Research Methods Courses

R. Eric Landrum
Boise State University
Randolph A. Smith
Kennesaw State University

Statistics and research methods courses provide the empirical foundation of psychology for undergraduate majors. In a review of college catalogs, Perlman and McCann (1999) found that 58% of undergraduate institutions offering a psychology major required statistics and 40% of undergraduate institutions required research methods. Thus, these courses receive considerable attention from faculty. For instance, Giesbrecht, Sell, Scialfa, Sandals, and Ehlers (1997) wrote extensively about the essential topics to be covered in these courses. Conners, Mccown, and Roskos-Ewoldsen (1998) articulated four major challenges in teaching statistics: motivating students, math anxiety, performance extremes, and long-term retention. The popularity and importance of these courses is also evidenced by a recent national conference specifically addressing these courses, "Finding Out: Best Practices in Teaching Research Methods and Statistics in Psychology," sponsored by the Society for the Teaching of Psychology, the National Institute on the Teaching of Psychology, and the Center for Excellence in Teaching and Learning at Kennesaw State University, held November, 2004 in Atlanta, Georgia.

The centerpiece of any course is the syllabus. McKeachie (2002) highlighted the role of the syllabus, which "helps students discover at the outset what is expected of them and gives them the security of knowing where they are going. At the same time, your wording of assignment topics can convey excitement and stimulate curiosity" (p. 17). The importance of the syllabus is acknowledged even in the American Psychological Association's (2002) Ethics Code, Section 7.03: "Psychologists take reasonable steps to ensure that course syllabi are accurate regarding the subject matter to be covered, bases for evaluating progress, and the nature of course experiences" (p. 1068). This chapter presents basic principles of good syllabus design specific to the content of statistics and research methods courses.

THE LITERATURE REGARDING GOOD SYLLABI

The syllabus is an essential means of communication with students. Madson, Melchert, and Whipp (2004) identified three main areas of the course syllabus: the course subject matter (i.e., course description), the course expectations and objectives, and the course evaluation procedures (such as major assignments and grading policies). Forsyth (2003) suggested that students key in on the assessment portion of the syllabus, but Raymark and Connor-Greene (2002) found that students often fail to remember details from the syllabus. Raymark and Connor-Greene suggested administering a brief take-home syllabus quiz to help students review the syllabus in greater detail.

Empirical work exists regarding the communication value of the syllabus as well as what students attend to in a syllabus, suggesting what constitutes a good syllabus. For example, Perrine, Lisle, and Tucker (1995) experimentally manipulated the presence or absence of a support statement in a syllabus and found that, generally, when an instructor offered outside-of-class help in a syllabus, students were more likely to seek support from the instructor. They also found an interesting effect of age in their study: Younger students were more reluctant to seek help when the offer of support was not in the syllabus, but older students asked for help regardless of whether the syllabus mentioned its availability.

Becker and Calhoon (1999) studied in great detail what introductory psychology students attend to and retain from the course syllabus. On the first day of class, introductory students rated the amount of attention they paid to particular items; students rated these items as "high attention": (a) examination or quiz dates, (b) due dates of assignments, (c) reading material covered by each exam or quiz, (d) grading procedures and policies, (e) type of exams or quizzes (e.g., multiple choice, essay), (f) dates and times of special events that students had to attend outside of class, (g) number of examinations and quizzes, (h) kind of assignments (e.g., readings, papers, presentations, projects); (i) class participation requirements, and (j) amount of work (e.g., amount of reading, length of assignments, number of assignments). Interestingly, Becker and Calhoon asked students again at the end of the semester about what they attended to in the syllabus, using a pre–post design. Of all the items exhibiting a significant change, the direction of the change was for students to give the items less attention at the end of the semester except for the following items (which increased in attention value at the end of the semester): (a) days, hours, and location of class meetings; (b) holidays; and (c) drop dates. Instructors can use this information to design effective course syllabi that emphasize those items deemed important by students. Additionally, the general questions provided in Table 4.1 (adapted from Forsyth, 2003) also provide a starting point for good syllabus design.

We devote the remaining portion of this chapter to presenting (a) specific ideas that apply to either a statistics or research methods course, (b)

TABLE 4.1

Types of Information Supplied to Students in a Typical Course Syllabus

Topic	*Issues to Consider*
Instructor	What is your name? What should students call you? Where is your office, and when do you hold office hours? Contact by e-mail, phone (office, home, or both—or never at home?)
Course Description and Goals	What are the overall course goals? How does this course contribute to the expected general education outcomes? What should students know when this course is complete? What should students be able to do at the completion of this course? How will this course change the participants?
Course Topics	What are the topics covered in this course? Why are these topics important? Is the order of presentation of topics important? What prerequisite information or coursework is recommended or required for successful course completion? Are class discussions graded?
Teaching and Learning Methods	What instructional methods will be used to teach the course material? Will this course make any unusual demands on students (writing intensive, use of technology, civic engagement, other projects)? What is the rationale for these teaching and learning methods?
Textbooks and Readings	What textbook(s) will be used? Will other reading assignments be made? Why were these texts and/or readings chosen? Are these primary or secondary sources?
Activities and Assignments	What types of learning activities and assignments will be made? Will papers be required? What is the purpose of these assignments? To what extent will class assignments influence course grade?
Grades	How will student progress be measured? Is the grading approach criterion-referenced or norm-referenced? How much is each activity and test worth? Can tests be dropped? Is the final exam cumulative? Due date policy listed?
Policies	What is the attendance policy (excused vs. unexcused absences)? Are there make-up tests? Can I bring guests to class? What is the policy regarding cell phones and other electronic devices? Can I use a laptop during an exam? Is there extra credit?

(continued)

TABLE 4.1 *(continued)*

Topic	*Issues to Consider*
Sources of Support	Will you hold review sessions prior to exams? Are the lecture notes available before or after class? What are the other sources of support on campus? Do you have teaching assistants? Will students have the opportunity to form study groups?
Calendar	When are assignments due? Due date policy? When should readings be completed by? When are tests? When and where is the final exam? What is the timetable for covering various topics? When is the deadline to withdraw from the class (with or without a "W" on the transcript)?
Academic Integrity Policy	What is the academic integrity policy for this university and for this class? Is there an honor code? What is plagiarism, and what types of penalties exist if plagiarism is detected?
Special Issues	What special considerations apply to this class? Should students be warned about materials or activities that they might find objectionable? Should students with special needs contact the professor? If a special situation arises during the semester, what is the best method of contacting the instructor?

Note. Adapted with permission from Forsyth (2003).

information about the frequency of particular practices, and (c) our recommendations for best practices. In some cases, we provide an example of that practice from an actual syllabus in use. For statistics courses, we found 35 course syllabi by doing a Google™ search on statistical methods and statistical methods and psychology; we also included all of the relevant course syllabi (2) from Project Syllabus (http://www.lemoyne.edu/OTRP/projectsyllabus.html). For research methods courses, we examined 25 syllabi based on the first author's presentation at the "Finding Out: Best Practices in Teaching Research Methods and Statistics in Psychology," at which conference attendees provided sample research methods syllabi; we also included all 9 relevant syllabi from the Project Syllabus Web site.

IDEAS SPECIFIC TO YOUR STATISTICS SYLLABUS

In this section we summarize the findings from our review of 37 statistics syllabi, and we also include our recommendations for best practices.

Course Objectives and Organization

Just over 15% of statistics syllabi indicated that there were specific computer lab assignments; these appeared to be a component of the lab section of the course. About 10% of syllabi also specifically mentioned prerequisites for the course and how these prerequisites were important preparation for the statistics course. Two of the syllabi also made reference to a two-course sequence in statistics. Also, one syllabus mentioned a team project for the course, and another syllabus referred to a computer exam that was required to be passed by the end of the course. We liked the course objectives offered by Gerard Saucier (2003) of the University of Oregon:

> This course is designed to help you gain the following: (a) The ability to understand and explain to others the statistical analyses in reports of social and behavioral science research; (b) preparation for learning about research methods, and about more advanced statistical methods; (c) the ability to identify the appropriate statistical procedure for many basic research situations and to carry out the necessary computations, by hand (for simple computations) or by computer (for more complex ones); and (d) further development of your quantitative and analytic thinking skills.

As for our best practice recommendations, we believe instructors should be as clear as possible in describing the course objectives to students and, furthermore, instructors need to demonstrate to students the connection between the course objectives and the tasks to be completed throughout the course. Making this connection explicit is particularly important in the course that Berk and Nanda (1998) described as the triple threat: super-boring, ultra-difficult, and most anxiety producing. We recommend that the course objectives and goals be as explicit and precise as possible. This practice is also helpful when course syllabi are reviewed during departmental assessment procedures as well as during institutional accreditation.

Use of Statistical Software

The most commonly listed statistical package in the statistics course was SPSS, specifically mentioned in syllabi about 13% of the time. Other software packages and programs mentioned included Excel, StatPro, SAS, and E-Stat. It may be the case that these (and other) statistical packages are indeed used more frequently but that they are not mentioned in the syllabus. This explanation of the use of SPSS by Anthony Greene (2005) of the University of Wisconsin-Milwaukee is quite appropriate:

> You will learn to use SPSS to solve all variety of statistical problems. This has three aims: First, you can focus much of your effort on the higher level aspects of statistics (experimental design, choosing the correct statistic, interpreting the outcomes of statistical tests, etc.). Second, you will be familiar with SPSS, which is widely used in research as well as in all variety of business contexts. Third, when you take your research methods course, SPSS is used.

Regarding supplemental SPSS books, the syllabi listed three different textbooks, with Kirkpatrick and Feeney (2001) most often mentioned.

We recommend that students have some hands-on experience with statistical calculations in a statistics course. This work does not necessarily have to be with a computer package, but students will have little understanding of statistics, in our opinion, if they do not get actively involved in the calculation of statistics. Exposure to a statistical package does not necessarily mean SPSS, although it is the most popular software option. Our best practices recommendation would be to have students involved in calculating actual statistics; the details of the process (actual data or problem sets; SPSS/Excel, by hand, or both) is left to the instructor to determine.

Calculators

The mention of the use of a calculator was popular in statistics syllabi—over 47% of syllabi specifically mentioned that the use of a simple calculator was recommended ("simple" meaning the ability to add, subtract, multiple, divide, and calculate square roots). Over 10% of the syllabi recommended a specific calculator, most commonly being the TI-83 (the Casio FX-300 was also mentioned). In fact, Stephen Marson (2003) from the University of North Carolina at Pembroke made this helpful recommendation in his syllabus: "One way of testing a calculator is by finding the square of –5. If your answer is –25, don't purchase the calculator."

We recommend that if you are going to have students calculate statistics by hand, then make some specific recommendations as to a calculator to use. Students often prefer to know what calculator the instructor uses because the instructor is often likely to be able to help students with the same model. Also, our best practices recommendation includes encouraging students to use a calculator for which the student also has the instruction manual. Mathematical symbols often differ on different models of calculators, and students need to be able to differentiate, for example, between the sample standard deviation and the population standard deviation. Be clear with students as to the type or level of calculator they will need (for example, a simple calculator, a statistics calculator, or a programmable calculator).

Textbooks

As you can imagine, there are many textbooks available for an undergraduate course in Statistics. In our review of 37 syllabi, we found 18 separate textbooks. From those 18, we found only 6 textbooks mentioned more than once, and those textbooks were (in alphabetical order) Aron and Aron (2003), Gravetter and Wallnau (2002), Hurlburt (2003), Jaccard and Becker (2002), McCall (2001), and Moore (2004). Two syllabi explicitly stated that there was not a textbook for the course; one syllabus also recommended for students was the *Publication Manual of the American Psychological Association* (2001).

As with any textbook adoption, instructors need to balance the rigor of the textbook with its user-friendliness and appeal to undergraduate students. This task is not a small one in the selection of a statistics textbook. There are a number of options available for textbook selection, such as books that cover both research methods and statistics (often designed for two-semester courses) or textbooks that focus on hand calculations or the comprehension of computer printouts. Other concerns may be to select a textbook that achieves departmental goals for the statistics course or selecting a book that best matches students' ability at a particular institution. Our best practices recommendation is for instructors to personally review textbook offerings from time to time to ensure that the text meets course goals and that the optimal textbook for the course, instructor, and students is actually in use. Required textbooks as listed on the syllabus should actually be required for success in the course.

IDEAS SPECIFIC TO YOUR RESEARCH METHODS SYLLABUS

Although there can certainly be overlap between the concepts discussed in a statistics course and the research methods course, here we present ideas from course syllabi that seem more specifically tied to the research methods course, including information about frequency of use and our opinion about best practices.

Course Objectives

Across all of the research methods syllabi reviewed, there was not a core or unified set of course objectives, although most syllabi did include course objectives or course goals. About 15% of the syllabi included specific information about separate lab sections of the course and articulated the objectives of those lab sections. Although syllabus authors exhibited great variability in their course objectives, some guidance does exist on a national level. APA's Task Force on Undergraduate Psychology Major Competencies (2002) identified 10 learning goals and outcomes for undergraduate psychology majors. Principle #2 concerns research methods in psychology, specifically that student will understand and apply basic research methods in psychology, including research design, data analysis, and interpretation. This goal can be articulated in a number of ways, which probably accounts for the variability in course objectives. For instance, this principle is reflected in the research methods syllabus of Bill Lammers (2004) at the University of Central Arkansas: "The purpose of the course is to (1) help students become critical consumers of information, (2) understand procedures that represent the most powerful ways of acquiring new knowledge, and (3) learn how to develop a research proposal (using APA format)."

Whatever the course goals are, we believe that best practices dictate that the course goals and objectives are communicated to students. Not only are students informed of the course goals, but they are also informed as to why these goals are important and how they are intended to be achieved in the course. Thinking about course goals and their incorporation into the syllabus can also help with assessment and accreditation efforts that are pervasive at our institutions today. Our best practice recommendation: include clearly defined course goals and review these goals with your students. We encourage instructors to design course goals in agreement with Principle #2 as recommended by the APA Task Force (2002).

Research Projects

In the syllabi we reviewed, actual research projects were not as common as we expected. About 30% of research methods syllabi described group projects, and about 20% of syllabi stated that students would conduct individual research projects. In some cases, students would complete multiple projects, such as one observational or field study and one experimental study. In about 15% of the syllabi, students wrote a research proposal but did not complete the actual experiment. In two cases, a two-course research methods sequence was available in which students wrote the proposal in one semester and completed the project in the subsequent semester. In about 12% of the syllabi, there was a class presentation of research results, and in some cases, this requirement entailed a poster presentation. A typical type of research project requirement that appeared in syllabi is that of Kelli Klebe (2000) of the University of Colorado at Colorado Springs:

> There are two course projects. Each project will be an empirical research study that you design, run, analyze, interpret, and write-up. The first project will be an observational project and the second project will be an experimental project. Each project needs to be non-threatening and follow ethical guidelines. The first project must be unobtrusive (no interaction with those you are observing). The second project must be an experiment that includes the manipulation of the independent variable.

What do we recommend as a best practice? There are institutional and departmental influences on what students and faculty can accomplish in a semester, as well as curricular constraints and the adequate preparation by prerequisite courses. Having said this, we recommend that students receive some "hands-on" experience with a research project. An ideal situation would be to finish a complete project that included data collection and analysis as well as the generation of an APA-formatted manuscript—this requires both careful design of course and syllabus. In situations where this type of requirement is not feasible, then some hands-on experience (writing the research proposal, collecting data on one's own class) is preferable to no actual research experience at all. Our best practices recommendation would be to collect and analyze real data, whether it be observational, quasiexperimental, or experimental, and to emulate the entire research process as much as possible.

Research Ethics

Because we believe that research ethics should be an essential component of any research methods course, we were surprised that research ethics is not mentioned more prominently in syllabi. Fewer than 20% of the syllabi mentioned research ethics. In the cases where ethics did appear, it sometimes appeared in the course objectives section, sometimes the syllabus specifically mentioned protection of human participants, once fraud was mentioned, and occasionally the IRB application was discussed. This situation is in interesting contrast to the general coverage of academic integrity or dishonesty issues, which occurred in the majority of research methods syllabi. It may be that the topics are covered in the course, but just not mentioned specifically in the course syllabus. We recommend that research ethics be given a more prominent place in the syllabus.

Our best practices recommendation: Instructors should at least mention research ethics in the course syllabus, and the presentation of ethical guidelines and procedures should be an integral part of every course in research methods. Perhaps instructors can integrate the topic of research ethics in the presentation of academic dishonesty/integrity policies in the syllabus, incorporating the ethical treatment of participants as one component of academic integrity for psychology majors.

APA Format

As part of the writing process in research methods, there is frequent mention of APA format and the *Publication Manual* (APA, 2001). About one-third of research methods syllabi made significant mention of APA format, listing it as a primary course goal or objective, including it as part of the course calendar, providing significant coverage of the details of APA format in the syllabus, or (in one case) mentioning APA format in the context of writing across the curriculum. Just over 50% of the instructors required the *Publication Manual* as a textbook, and another 20% recommended it to students. Across the syllabi, there were listings for four other writing guides (either required or recommended), with the most commonly cited being R. L. Rosnow and M. Rosnow (2001) and Perrin (2004). We appreciate the APA-writing philosophy expressed by Connie Wolfe (2004) now at Muhlenberg College:

> You may have been instructed in previous psychology classes to conform to some subset of the APA guidelines for writing your papers. Welcome to what is probably your first exposure to APA style in all its glory! This will probably seem very tedious (okay, it IS very tedious), but it is important. One goal of this course is to teach you how to write a clear, organized, APA-style research paper. APA style will haunt you if you continue on to graduate school in psychology. Even if you don't, however, you will have to follow some sort of guidelines no matter where you go or what you write in the future. At the very least, you will certainly always need to be clear, organized and grammatically accurate in your writing. Consider learning APA style to be a vigorous exercise in writing well. I strongly encourage you to begin acquainting yourself with the *Publication Manual* and the rules in it now. (Lesson one: forget everything you learned about "MLA" style). Trying to make your paper conform to APA guidelines the night before it is due will may you crazy and result, probably, in a not-so-good paper.

We believe that learning to master APA format results in valuable, transferable skills to either a career with a bachelor's degree or the pursuit of graduate education. The ability to learn to follow instructions may ultimately be more important than the particular instructions themselves. Our best practices recommendation: Introduce APA format into your research methods classes to the extent that you feel comfortable, and hold students to a high (but fair) standard in submitting their work in APA format—make sure you mention APA format in your syllabus to draw attention to its importance. Multiple opportunities to practice APA format during the course will allow you to shape students into the proper use of APA format.

Textbooks

The opportunity to review 34 research methods syllabi in detail allowed us to see how instructors design and organize their courses. An important part of this course design is the selection of a textbook. Not surprisingly, no consensus textbook choice emerged from our review of the syllabi. In fact, 19 different textbooks were identified across the 34 syllabi. No book was mentioned more than 4 times out of the 19 different texts. The most commonly mentioned textbooks, each being mentioned at least twice, included (in alphabetical order) Bordens and Abbott (2005); Cozby (2000); Goodwin (2003); Mitchell and Jolley (2004); Shaughnessy, Zechmeister, and Zechmeister (2003); and Smith and Davis (2003). Additionally, some research methods syllabi also recommended a SPSS guide; 5 different guides were recommended overall. The most commonly recommended guides were Cronk (2004) and Pavkov and Pierce (2003).

Selecting a textbook is both an important and personal decision for most instructors. Textbook selection is affected by a number of factors, both general to the institution and specific to the instructor. We cannot recommend one textbook to all instructors, because indeed one size does not fit all, and obviously the market can bear multiple successful books on the topic of research methods. Our best practices recommendation: Carefully consider all of your textbook options, and select the textbook that best meets both your needs and the needs of your students. Some textbooks focus exclusively on research methods, whereas others address both statistics and research methods—be sure to select the textbook that best matches the format of your course. If a textbook selection does not work out as hoped, do not be afraid to change—there are obviously many good choices available.

CONCLUSIONS

Statistics and research methods courses form the methodological core for many (or most) undergraduate psychology departments. After examining course syllabi from around the country, it is clear that instructors tend to deal with a core set of issues. Interestingly, faculty use a variety of approaches to achieve course goals. Whether through searching the Internet or attending specialty area conferences or using Project Syllabus, we recommend that our colleagues not only share syllabus information with others, but also seek out syllabus information. We came across many good ideas in our review of syllabi, and some of these good ideas were not course-specific (exam make-up policies, attendance policies, use of teaching assistants, disability statements, extra credit policies, issues of academic integrity, time management strategies), so we did not include them in this chapter.

Whereas we have identified what we believe are some best practices, truly best practices are situation specific and must be designed to meet the needs of the course, department, university, and student. One resource that makes this process easier is Project Syllabus, sponsored by the Society for the Teaching of Psychology (Division Two of APA). Project Syllabus resources are located online at http://www.lemoyne.edu/OTRP/projectsyllabus.html. We recommend that faculty access and take advantage of the accumulated wealth of knowledge available from syllabi and colleagues nationwide. Additionally, we were surprised with the relatively low number of research methods and statistics syllabi available from Project Syllabus given the prevalence and importance of these courses. We also recommend that more instructors consider submitting their course syllabi to this vital resource.

REFERENCES

American Psychological Association. (2001). *Publication manual of the American Psychological Association* (5th ed.). Washington, DC: Author.

American Psychological Association. (2002). Ethical principles of psychologists and code of conduct. *American Psychologist, 57,* 1060–1073.

Aron, A., & Aron, E. (2003). *Statistics for psychology* (3rd ed.). Upper Saddle River, NJ: Prentice Hall.

Becker, A. H., & Calhoon, S. K. (1999). What introductory psychology students attend to on a course syllabus. *Teaching of Psychology, 26*, 6–11.

Berk, R. A., & Nanda, J. P. (1998). Effects of jocular instructional methods on attitudes, anxiety, and achievement in statistics courses. *Humor: International Journal of Humor Research, 11*, 383–409.

Bordens, K. S., & Abbott, B. B. (2005). *Research design and methods: A process approach* (6th ed.). Boston: McGraw-Hill.

Conners, F. A., Mccown, S. M., & Roskos-Ewoldsen, B. (1998). Unique challenges in teaching undergraduate statistics. *Teaching of Psychology, 25*, 40–42.

Cozby, P. C. (2000). *Methods in behavioral research* (7th ed.). Mountain View, CA: Mayfield.

Cronk, B. (2004). *How to use SPSS* (3rd ed.). Glendale, CA: Pyrczak.

Forsyth, D. R. (2003). *The professor's guide to teaching: Psychological principles and practices*. Washington, DC: American Psychological Association.

Giesbrecht, N., Sell, Y., Scialfa, C., Sandals, L., & Ehlers, P. (1997). Essential topics in introductory statistics and methodology courses. *Teaching of Psychology*, 24, 242–246.

Goodwin, C. J. (2003). *Research in psychology: Methods and design* (3rd ed.). Hoboken, NJ: Wiley.

Gravetter, F., & Wallnau, L. (2002). *Essential statistics for the behavioral sciences* (4th ed.). Pacific Grove, CA: Brooks/Cole.

Greene, A. (2005). *Psychological statistics*. Unpublished syllabus, University of Wisconsin – Milwaukee.

Hurlburt, R. T. (2003). *Comprehending behavioral statistics* (3rd ed.). Belmont, CA: Wadsworth.

Jaccard, J., & Becker, M. A. (2002). *Statistics for the behavioral sciences* (4th ed.). Belmont, CA: Wadsworth.

Klebe, K. (2000). *Introduction to psychological research and measurement*. Unpublished syllabus, University of Colorado at Colorado Springs.

Kirkpatrick, L. A., & Feeney, B. C. (2005). *A simple guide to SPSS for Windows*. Belmont, CA: Wadsworth.

Lammers, B. (2004). *Research methods*. Unpublished syllabus, University of Central Arkansas, Conway, AR.

Madson, M. B., Melchert, T. P., & Whipp, J. L. (2004). Assessing student exposure to and use of computer technologies through an examination of course syllabi. *Assessment & Evaluation in Higher Education, 29*, 549–561.

Marson, S. M. (2003). *Social statistics syllabus*. Retrieved July 25, 2005, from http://www.uncp.edu/home/marson/360_summer.html

McCall, R. B. (2001). *Fundamental statistics for behavioral sciences* (8th ed.). Belmont, CA: Wadsworth.

McKeachie, W. J. (2002). *McKeachie's teaching tips: Strategies, research, and theory for college and university teachers* (11th ed.). Boston: Houghton-Mifflin.

Mitchell, M., & Jolley, J. (2004). *Research design explained* (5th ed.). Belmont, CA: Wadsworth.

Moore, D. S. (2004). *Introductory statistics* (5th ed.). New York: W.H. Freeman.

Pavkov, T. W., & Pierce, K. A. (2003). *Ready, set, go! A student guide to SPSS 11.0 for Windows*. Toronto: Mayfield.

Perlman, B., & McCann, L. I. (1999). The structure of the psychology undergraduate curriculum. *Teaching of Psychology, 26*, 171–176.

Perrin, R. (2004). *Pocket guide to APA style*. Boston: Houghton-Mifflin.

Perrine, R. M., Lisle, J., & Tucker, D. L. (1995). Effects of a syllabus offer of help, student age, and class size on college students' willingness to seek support from faculty. *Journal of Experimental Education, 64*, 41–52.

Raymark, P. H., & Connor-Greene, P. A. (2002). The syllabus quiz. *Teaching of Psychology, 29*, 286–288.

Rosnow, R. L., & Rosnow, M. (2003). *Writing papers in psychology* (6th ed.). Belmont, CA: Wadsworth.

Saucier, G. (2003). *Statistics*. Retrieved July 25, 2005, from http://darkwing.uoregon.edu/~gsaucier/Psych_302_syllabus.htm

Shaughnessy, J. J., Zechmeister, E. B., & Zechmeister, J. S. (2003). *Research methods in psychology* (6th ed.). New York: McGraw-Hill.

Smith, R. A., & Davis, S. F. (2003). *The psychologist as detective: An introduction to conducting research in psychology* (3rd ed.). Upper Saddle River, NJ: Prentice Hall.

Task Force on Undergraduate Psychology Major Competencies. (2002, March). *Undergraduate psychology major learning goals and outcomes: A report*. Washington, DC: American Psychological Association. Available at http://www.apa.org/ed/pcue/taskforcereport2.pdf

Wolfe, C. (2004). *Research methods & statistics*. Unpublished syllabus, Muhlenberg College, Allentown, PA.

Chapter **5**

Assessing Students' Research Ideas

Paul C. Smith
Alverno College

I teach psychology at Alverno College, a private women's liberal arts college where experimental psychology is a junior-level course required for majors. The students' main task in this course is to design and conduct a semester-long independent "true" experimental research project that tests a causal hypothesis and is connected to published theoretical and research literature. On the first day of class, students typically share tentative ideas for their projects. Despite having had extensive preparation in earlier coursework, many students have an extremely difficult time coming up with usable ideas. Some students fail to come up with any idea at all. But it is at least as common for a student—even one who is generally strong academically—to confidently propose a research idea that is inappropriate in one of a few distinct ways. In the semesters that I taught experimental psychology or helped students develop their research proposals, I noticed that students reliably gravitated toward certain kinds of inappropriate project ideas, even after our most careful preparations.

A teacher's normal response to this kind of situation is to develop exercises and materials to help guide the students to proper ideas. Certainly, with enough preparation, faculty can head off those inappropriate ideas, and in fact, students do eventually produce good projects. But in their haste to get students to produce a methodologically sound and substantively interesting project, faculty should not miss the opportunity to learn something about students' thinking from the project ideas they bring to the class. I am curious about my students' difficulties, not just because I want them to come to the class with better ideas, but because I think the ideas that they do bring to the class are valuable assessment data. Taken seriously, the patterns of inappropriate ideas can reveal a lot, not just about gaps in students' preparation, but also about their beliefs, skills, and interests in psychology. The tendency toward certain kinds of inappropriate projects may be a symptom of some deeper conceptual and ability issues, ones that might have an impact elsewhere in the curriculum and in their work after graduation.

THE NATURE OF THE PROBLEM

My department's curriculum is fairly typical in its efforts to prepare students for their upper level research coursework. Students in experimental psychology have taken an introductory course, courses in lifespan development and abnormal psychology, and generally at least one upper level theory-based psychology course. This is a very typical set of required core coursework (Perlman & McCann, 1999). Our students also have completed an introductory research methods course as well as introductory probability and statistics. This preparatory methodology sequence is similar to that required at 32% of the institutions in a national sample examined by Messer, Griggs, and Jackson (1999).

Despite this preparation, both in psychology's content and in its methods, many students come to the first day of the experimental course with research ideas that are inappropriate for the course outcomes. These unsuitable ideas tend to take one of a few clear forms, which I call the *noncausal project*, the *breakthrough project*, and the *"so what?" project*.

The Noncausal Project

In our experimental psychology course, it is not unusual for most of the students who have an idea in mind when they come to the first day of class to propose ideas that are noncausal in nature. Instead of describing some experimental manipulation and a way of looking for effects of that manipulation, many students proposed comparisons between two existing groups, usually normal/abnormal contrasts or gender contrasts. For example, a student may tell us that she wants to compare alcoholics' and nonalcoholics' self esteem or compare the problem-solving styles of men and women. On more than one occasion, students confidently proposed existing group comparisons immediately after I had carefully explained to the previous student why that was inappropriate for this course.

The problem does not seem to be a lack of attention or motivation or simply that the students are unaware of the course requirements. Instead, students often seem driven away from true experimental projects and almost willfully resist efforts to move them toward experimental projects. After one student proposed a gender comparison, two faculty members worked with her to help her to develop a related causal hypothesis. Once we had agreed on an experimental manipulation, we were confident enough to let her continue her work independently. At the students' poster session at the end of the semester, we discovered that despite having included the manipulation in her design, her conclusions ignored it entirely and instead were only about the gender difference that she had originally proposed. I asked her why she had included the manipulation, and she confidently replied that it was because she needed it for the study to be a true experiment. Her commitment to the original idea was so strong that she had convinced herself that the experimental requirement was just a hoop that we expected her to jump through.

To get a sense of the extent to which students in general tend toward existing group comparisons, I examined undergraduates' articles published in the *Psi Chi Journal of Undergraduate Research* from Volume 1, Issue 1 (Spring 1996) to Volume 9, Issue 2 (Summer 2004). I found that about 39% of the studies published there included a manipulated variable, whereas about 48% included existing group comparisons (many included both). Roughly 22% of the studies included gender comparisons, and those gender comparisons formed the main hypothesis for about 12% of all studies. Of course the published projects are not a representative sample of all students' project ideas, and there is no requirement that these studies have a true experimental design, but it seems unlikely that true experiments are underrepresented here when compared with students' research ideas in general. By contrast, a scan of articles in *Psychological Science* from the same time period showed that about three quarters of those studies included manipulated variables, and in the 10 issues I examined, no studies focused on gender comparisons (interestingly though, the *Psychological Science* studies did include more developmental comparisons among age groups than did the undergraduates' studies).

Why, when left to their own devices, do so many students propose existing group comparisons, rather than causal questions? The most obvious answer is that these existing group comparisons are what spark students' curiosity. It seems that the students who propose these projects and cling to them even in the face of feedback from faculty are relatively uninterested in situational and environmental causes of behavior. Instead, their interests seem more focused on how groups of people—typically men and women—are different from one another.

The Breakthrough Project

Occasionally, students come to the class with a clear idea in mind for a project that does include a causal hypothesis but that is inappropriate for other reasons. For example, some students come with *breakthrough projects*: research ideas that are intended to challenge certain basic tenets of psychology or to push the boundaries of the discipline. Occasionally, students want to use their experimental psychology projects to show that their paranormal beliefs were true. For example, a few years ago, a student designed a study of "the power of intercessionary prayer," and others have proposed ideas for testing meditation or "aromatherapy."

My objection here is not that certain topics are out of bounds simply because they are not mainstream psychology. The problem is with a notion of scientific breakthroughs. The student's motivation is a desire to defend her belief in the face of the psychological literature, and she believes that her project can provide her with that defense.

In his wonderful book, *How to Think Straight About Psychology* (2004), Stanovich devoted one chapter to a discussion of naive notions about the importance of breakthroughs to science, and suggested that the public is so taken with the idea of scientific breakthroughs, in part, because the *breakthrough* label is such a common part of news reports about science. The story of the lone researcher working against a unified and dogmatic scientific consensus is an established part of popular culture but it misrepresents how science progresses. Faculty are presumably "immune" to these sensationalist portrayals of research because, unlike most students, we have had the experience of reading large collections of sometimes conflicting research articles about some topic and of trying to judge where the evidence converges. As a result, we come to deeply understand that there is always ambiguity in the collection of findings on a particular topic, and that change happens as a result of what Stanovich (2004) calls "the gradual-synthesis model" (p.116), slowly tracking changes in the preponderance of the evidence. The beginning psychology student does not have that experience and may assume that a single breakthrough project is a perfectly reasonable challenge to the discipline's stance on a question.

Occasionally, of course, there really are valuable studies that do have a dramatic impact on psychology. Hock's (2005) *40 Studies that Changed Psychology* presents examples, with commentary about the theoretical background and the impact of each. Hock chose his 40 studies explicitly because of their impact on the direction of psychology: Each caused people to think differently about an area of psychology and inspired considerable further research. However, Hock's 40 studies challenge, rather than reinforce, the popular notion of a scientific breakthrough. Each is tightly tied to earlier research and theory, and the studies had their impact only because they were so well connected to the existing literature. By contrast, the typical student breakthrough project is an attempt to reject or ignore the literature, and, as a result, is all but certain to find itself ignored. Not surprisingly, I found no breakthrough projects in my examination of the *Psi Chi Journal of Undergraduate Research*. In proposing a breakthrough project, a student reveals a particular naive notion of how individual research projects fit into the larger literature and a sense that the value of research is in the extent to which it supports her prior beliefs (Smith, 2000).

The "So What?" Project

At the other end of the spectrum from breakthrough projects are "so what?" projects—those that answer only the most trivial questions.

Classic examples include "the effects of music on studying" (studying with and without music playing or with different kinds of music) and "the effects of color on memory" (studying text on paper of different colors). Students can turn these ideas into meaningful studies, for example, by applying a theory of attention to make an informed prediction about what it is about music that interferes with studying and then devising a test of that specific prediction. However, in their rush to help the student identify a meaningful study, faculty should not forget to wonder about what it was in the student's thinking that led her to the original "so what?" idea. When a student proposes a "so what?" project, she is probably simply trying to meet the basic course requirements and does not value the project as an opportunity to learn something about psychology. It is tempting to make a dispositional attribution here that the student simply is not a hard worker. However, if she does not properly understand the task of research conceptualization, a motivated and hard-working student may nonetheless choose to treat the experimental project requirement as a mere hoop to jump through. Faculty can make themselves aware of the sources of students' difficulties through an analysis of students' understandings of the research conceptualization task.

UNDERSTANDING STUDENTS' UNDERSTANDINGS OF RESEARCH CONCEPTUALIZATION

Assessing for Student Conceptions

At Alverno College, assessment is not simply something done in order to decide whether a student has learned the material that has been covered. Instead assessment is an integral part of ongoing student learning and teaching (Alverno College, 2001), informs ongoing instruction, guides the design of instruction and curricula, and validates programs, instruction, and curricula. We do not use it only to establish that a student has developed the abilities that we expect of her.

Assessment is most useful to learning and teaching when the faculty design assessments to be diagnostic, in order to reveal the specific conceptions that students use in their performances. In *Knowing What Students Know: The Science and Design of Educational Assessment* (National Research Council, 2001), the Committee on the Foundations of Assessment described an "assessment triangle," relating observation, interpretation, and cognition. This model focused on the fact that instructors use their observations of students' performances to make inferences about the quality of the students' skills and knowledge. Those inferences depend on instructors' models of learning in the domain, particularly the assumptions about the skills and knowledge that students need in order to perform the task. In turn, those inferences direct teaching. They not only point to gaps in students' skills and knowledge but also suggest misconceptions that the teacher must actively address. In short, what matters in assessment is not simply whether students can do what instructors ask them to do, but also how instructors interpret students' performances in order to make inferences about students' thinking about the task.

The research conceptualization task in the experimental psychology course provides a special opportunity to look into students' thinking about the nature of knowledge in the field of psychology. Students come to the experimental psychology course with widely different levels of preparation. Some have good ideas in hand already, perhaps needing the instructor's input only on measures and on the mechanics of finding participants and arranging data-gathering sessions. At the other end of the spectrum, some students have no ideas at all for their projects. This disparity certainly has many causes, and we will certainly never know all of the reasons why some students find the conceptualization task easy and others do not. However, consistent and persistent problems like the noncausal, breakthrough, and the "so what?" project ideas discussed earlier strongly suggest that students have certain specific misconceptions about psychology, research, or both. An understanding of these misconceptions can guide faculty in the development of better methods to help students to find appropriate research projects.

Noncausal Ideas Suggest Alternative Conceptions of Psychology

When students persist in noncausal ideas, they reveal a particularly important misconception about psychology. Psychologists take the philosophical position that people do what they do, at least in part, for reasons having to do with their specific interactions with the environment. This *environmental interactions* position raises the causal questions that constitute the domain of experimental psychology. But there are two competing philosophical positions that students (and the lay public in general) may hold. These are the *essentialist position* that people do what they do because they are the kinds of people who do that kind of thing and the *strong free-will position* that people do what they do because they freely choose to do those things. In contrast to the environmental interactions position, the essentialist and strong free-will positions do not support the curiosity about causal questions necessary for experimental research conceptualization.

A student who holds an essentialist position explains behavior by referring to a person's internal essential nature and therefore believes that the purpose of research in psychology is only to uncover that nature. For example, the essentialist believes that if a person abuses drugs, it is because that person is a drug abuser (or "is the kind of person who abuses drugs"). As a result, her research questions will tend to be descriptive ones about other ways that such people differ from the kinds of people who do not abuse drugs, rather than causal ones about why some people abuse drugs and others do not. It was during work with a student in my senior seminar course that I first began to suspect that some of my students' difficulties arose from this essentialist position. In that course, students develop theoretical explanations of some behavior they find interesting. My student's career interest was in working to prevent child abuse, so we agreed that her project would be about why some adults abuse children. But her first draft had nothing on that question and instead was a listing of statistics on the prevalence of abuse. We discussed the project again until she assured me that she understood, but her second draft also had no attempt to explain why an adult would abuse a child. She had become very familiar with statistics about the prevalence of child abuse and with organizations working to help abused children, but her assumption about why some adults abuse children had remained untouched by her education: It is because they are child abusers. It apparently did not occur to her that the abusers' behavior could be the target of any other sort of explanation.

This essentialist philosophical position helps to explain why students so often propose comparisons of existing groups for their project ideas. The essentialist position may also be behind the persistence of the confusion between naming and explaining, as reflected for example in the belief that labeling a person *schizophrenic* explains why that person hears nonexistent voices (Brewer et al., 1993). When a student uses a label as though it were an explanation or has a persistent difficulty coming up with research ideas that do not simply involve existing group comparisons, she may be revealing a pervasive tacit commitment to an essentialist position on psychology rather than a very specific lack of understanding or preparation.

The strong free-will position also causes student difficulties in research conceptualization. The student who holds this position believes that human beings do what they do simply because they freely choose to do so, and so she believes that she already has the only explanation necessary for normal behaviors. What can she make of the requirement that she design a study to answer a question about the cause of some specific behavior? One possibility is that she will believe that the only interesting questions have to do with the behavior of persons whom she believes lack free will—perhaps persons suffering from particular psychological problems, or children. I suspect that this position is partly why so many students propose normal/abnormal comparisons and are so disappointed to find out that for practical and ethical reasons, those comparisons are not appropriate as project ideas for the experimental psychology course.

Sometimes students explicitly cite the strong free-will perspective to deny the very possibility of behavioral science research on human beings, claiming that because human beings have free will, it is impossible to predict or control the behavior of individuals. There must be something very compelling about the notion of free will that adults can make such a claim when it is clearly contradicted by not only a long history of successful research and application, but also by almost all of our day-to-day experience (for example, the ability to predict that the driver in the oncoming lane will remain in that lane).

ADDRESSING THESE CONCEPTIONS: SUGGESTIONS FOR FACULTY

As noted earlier, this analysis of students' difficulties suggests a set of practices for faculty who want to guide students toward appropriate experimental research projects. Of course, many students' difficulties have nothing to do with these alternative conceptions and instead reflect a simple lack of preparation in the discipline. I propose that the suggestions I offer here supplement and not replace the more typical activities that faculty already use to help students to conceptualize their research projects. However, it is important that faculty not merely dwell on what students do not know, but also address what they know that is not so.

Suggestion 1: Contrast Essentialism and Free Will with the Environmental Interactions Position

The essentialist and strong free-will positions form part of a naive intuitive psychology. Students who operate from those positions are almost certainly unaware that they are doing so or that there are alternatives. At the same time, faculty do not generally think about students' difficulties in these broad terms, and instead, are likely to attribute them to a lack of effort or curiosity, or to a lack of more specific knowledge about psychology and research methods.

Faculty can take a good first step toward addressing these difficulties by making these tacit philosophical positions explicit. Research methods textbooks typically address causality, but do not explicitly discuss it as a philosophical position about why people do what they do or contrast it with alternative positions. That task is left to the instructor. At my institution, we have the luxury of an introductory research methods course before the experimental psychology course. That introductory course gives us the opportunity to prepare students well in advance of when they have to find their own research ideas, with discussions of the different philosophical positions as well as of multiple causation and other important topics forming the background for psychological research design. Those discussions should help the students reframe their understandings of the causes of behaviors in a way more conducive to research conceptualization, and at the same time, help the instructor to reframe her understanding of the students' difficulties.

Suggestion 2: Use Research Examples That Address Everyday Behaviors Among Normal Populations

Methodology courses tend to be about the "nuts and bolts" of research and statistical methods, and faculty generally assume that students will understand that those methods apply equally to the whole range of populations and psychological phenomena. This is not a safe assumption. Students who hold a strong free-will perspective will come to the research methods course believing that research methods are not necessary for answering questions about the causes of normal behavior of normal people, because free will already answers those questions. If instructors in the methodology courses use research examples that involve only abnormal populations or unusual behaviors, those students will find their misconception reinforced and as a result, restrict their search for experimental research project ideas to those involving abnormal populations or unusual behaviors and situations. In order to address this problem, instructors should make a specific effort to include among their examples research projects addressing normal everyday behaviors of normal populations.

Furthermore, instructors should explicitly point out that these examples address normal behaviors, rather than just assume that students will spontaneously notice. Students have a number of methods for defending their misconceptions from conflicts with the subject matter of the course (Bain, 2004; Chinn & Brewer, 1993, 1998). These methods include simply failing to notice a conflict with instruction as well as reinterpreting a conflict in some nonproductive way. Unless the instructor is completely explicit about the point, students who assume that free will is the only explanation needed for normal behaviors among normal persons will find a way to miss the lessons intended by the use of studies of normal behavior.

Suggestion 3: Discuss Inconsistencies in the Literature

As noted earlier, students who propose breakthrough projects are likely to believe the popular misconception that progress in research is largely a matter of discrete "leaps" driven by solitary researchers who question an otherwise monolithic scientific establishment. Faculty can challenge this misconception by explicitly drawing attention to the inconsistencies found in the literature and holding critical discussions of those inconsistencies with the class. With the growing popularity of meta-analysis, it is easier than ever to find articles that identify inconsistencies and that model the process of working through those inconsistencies to make well-supported recommendations. Student participation in discussion of these inconsistencies will serve a double purpose, addressing the popular misconceptions about the nature of progress in research, and pointing students who do not already have projects in mind to the most fertile ground for specific research ideas.

Suggestion 4: Brainstorm Using Conceptual Categories That Lead to Research Ideas

A student who proposes a "so what?" project or who has trouble coming up with a research idea at all may be just overwhelmed by her prior coursework. Faculty design psychology curricula, using one set of conceptual categories—the content areas such as social, personality, physiological, and cognitive—to organize the content of the discipline of psychology. These categories are not the only way to organize psychology's content, though, and they are not the categories that researchers use to find research ideas. There is no reason to think that these categories are any more useful for students faced with that task. Other conceptual categories of the discipline of psychology lead to research ideas far more directly than do the categories reflected in the titles of undergraduate courses. Students who do not have ideas in hand will find it helpful to brainstorm in terms of phenomena (e.g., the dramatic pace of the acquisition of vocabulary in the first months of learning to speak; Woodward, Markman & Fitzsimmons, 1994), terms used in the field to describe those phenomena (e.g., "fundamental attribution error"; Ross, 1977), theories and models (e.g., the modal model of memory; Anderson, 1999), practices in professional psychology (e.g., the use of psychological debriefing for trauma victims; McNally, Bryant, & Ehlers, 2003), and ongoing disputes in the field (e.g., "Are there factors in childhood experience that have a strong impact on adult sexual orientation?"; P. Cameron & K. Cameron, 1995). These conceptual categories are the ones that lead students' curiosity toward significant and researchable questions. A classroom discussion using these categories should help the students who cannot find a meaningful idea to break out of her nonproductive conceptual categories.

Suggestion 5: Head Off the Inappropriate Ideas in Advance

Besides the cognitive defenses, students also bring affective responses with their research ideas. A student who comes to the class with a breakthrough project is likely to be excited about her idea. Similarly, a student who comes with a "so what?" project may believe that her idea is a substantive one and be driven by curiosity about the results. Although it is useful for the instructor to know about students who bring those kinds of ideas to the course, it is emotionally difficult for a student to be put on the spot to identify an idea and then be told in public that her idea is inappropriate. Instructors who would like to learn about students' misconceptions from the ideas that they bring to the class face a difficult balancing act between the value of what we can learn from students' inappropriate project ideas and the need to be sensitive to students' emotional commitments to those ideas.

One way for faculty to be sensitive to students while directing them away from inappropriate project ideas is to point out in advance of hearing students' ideas that it is not at all unusual for students to come to the course with project ideas that are inappropriate. Part of the value of a psychology education is that it teaches students to wonder about environmental causes of behavior. The students who come to the course without that kind of directed curiosity are the ones who will benefit the most from the course. If faculty prime students in advance to think of critiques of their proposals as planned and valuable learning experiences, those students are far less likely to have negative emotional responses to those critiques.

CONCLUSION

There is certainly a wide variety of reasons that students have difficulties in finding independent experimental research ideas. Although faculty may find it easy to simply assume that students' difficulties come from a lack of knowledge of psychology or from a lack of effort, an analysis of the inappropriate ideas that they bring to the class suggests that certain misconceptions about psychology also contribute to the problem. Patterns of inappropriate research proposals suggest that faculty need to explicitly address students' basic assumptions about the causes of behavior in order to guide students to the kind of question-asking that leads to appropriate experimental research ideas.

ACKNOWLEDGMENTS

This work was originally developed with the support of an Alverno College Faculty Fellowship in the summer of 2001. I would like to thank my colleagues in that fellowship work, Cynthia Gray and Brenda Kilpatrick, whose comments raised my curiosity about this question. I would also like to thank the other members of the Department of Psychology, particularly Kris Vasquez and Joyce Tang Boyland, the current research methods and experimental psychology instructors, for their help in developing this work.

REFERENCES

Alverno College. (2001). *Student assessment-as-learning at Alverno College*. Retrieved July 8, 2005, from http://depts.alverno.edu/saal/

Anderson, J. R. (1999). *Cognitive psychology and its implications* (5th ed.). New York: Worth Publishers.

Bain, K. (2004). *What the best college teachers do*. Cambridge, MA: Harvard University Press.

Brewer, C. L., Hopkins, J. R., Kimble, G. A., Matlin, M. W., McCann, L. I., McNeil, O. V., Nodine, B. F., Quinn, V. N., & Saundra. (1993). Curriculum. In T. V. McGovern, (Ed.), *Handbook for enhancing undergraduate education in psychology* (pp. 161–182). Washington DC: American Psychological Association.

Cameron, P., & Cameron, K. (1995). Does incest cause homosexuality? *Psychological Reports, 76,* 611–621.

Chinn, C. A., & Brewer, W. F. (1993). The role of anomalous data in knowledge acquisition: A theoretical framework and implications for science instruction. *Review of Educational Research, 63,* 1–49.

Chinn, C. A., & Brewer, W. F. (1998). An empirical test of a taxonomy of responses to anomalous data in science. *Journal of Research in Science Teaching, 35,* 623–654.

Hock, R. R. (2005). *Forty studies that changed psychology: Explorations into the history of psychological research.* Upper Saddle River, NJ: Pearson Prentice-Hall.

McNally, R. J., Bryant, R. A., & Ehlers, A. (2003). Does early psychological intervention promote recovery from posttraumatic stress? *Psychological Science in the Public Interest, 4,* 45–79.

Messer, W. S., Griggs, R. A., & Jackson, S. L. (1999). A national survey of undergraduate psychology degree options and major requirements. *Teaching of Psychology, 26,* 164–171.

National Research Council. (2001). *Knowing what students know: The science and design of educational assessment.* Washington DC: National Academy Press.

Perlman, B., & McCann, L. I. (1999). The structure of the psychology undergraduate curriculum. *Teaching of Psychology, 26,* 171–176.

Ross, L. (1977). The intuitive psychologist and his shortcomings: Distortions in the attribution process. In L. Berkowitz (Ed.), *Advances in experimental social psychology* (Vol. 10, pp. 173–220). New York: Academic Press.

Smith, P. C. (2000). Promoting belief change by encouraging evaluation of prior beliefs. (Doctoral dissertation, University of Wisconsin at Milwaukee, 2000). *Dissertation Abstracts International Section A: Humanities & Social Sciences. 61*(3-A), 886.

Stanovich, K. E. (2004). *How to think straight about psychology* (7th ed.). New York: Harper Collins.

Woodward, A. L., Markman, E. M., & Fitzsimmons, C. M. (1994). Rapid word learning in 13- and 18-month-olds. *Developmental Psychology, 30,* 553–566.

Part III

Approaches to Teaching Statistics

Chapter 6

Designing Effective Examples and Problems for Teaching Statistics

Stephen L. Chew
Samford University

The use of examples to help students learn is common to virtually all teaching methods in all disciplines, but the importance of examples is especially clear in teaching statistics. Worked examples, usually a prototypical problem followed by the solution, are a universal feature of statistics textbooks, and it is hard to imagine a teacher who does not use examples to illustrate statistical methods.

Given the obvious importance of examples for teaching statistics, one might think that teachers would carefully design examples and assess their impact on student learning. In my experience, however, teachers typically create examples haphazardly or simply take them from the textbook. Teachers rarely assess the effectiveness of their examples for student learning. They seem to believe that as long as the example is accurate, it will be useful, and as long as the example is explained, students will automatically learn from it. Research on how to design and use examples most effectively for student learning indicates that these beliefs are simplistic. The design and use of examples is neither simple nor straightforward, and although the proper use of examples can be a powerful learning tool, their inappropriate use can impede the development of student understanding.

In this chapter I first review the existing research literature on creating and using examples effectively, with particular reference to their use for teaching statistics. I also discuss what students learn from examples. Finally, I propose a tentative model of how to design and use examples effectively that should help statistics teachers create good examples more systematically.

DEFINING AN EXAMPLE

For this chapter, I am defining an example as any specific instance, illustration, demonstration or activity that is representative of a concept. This definition is broad and includes activities, exercises, guided problem solving, and problem sets. A problem either worked in class by the instructor or assigned to students to solve would qualify as an example. Activities designed to demonstrate a statistical concept, such as generating sampling distributions, would also qualify.

Examples can be of two broad types. First, an example can be a prototypical exemplar or instantiation of a concept. For example, the two-group experimental design with separate experimental and control groups is a classic example for which one would use a *t* test with independent samples to analyze the data. An example can also be an analogy, where the example resembles the concept on some key dimension. For example, statistical hypothesis testing is similar to the decision a jury makes in a criminal trial. The defendant, who is either guilty or not guilty, can be found guilty or not guilty by the jury, leading to four possible outcomes, two of which are correct and two of which are erroneous.

THE EFFECTIVENESS OF EXAMPLES FOR LEARNING

The fundamental question about examples is whether they are effective for enhancing student learning. Research consistently finds that students rely heavily on examples and even prefer to study examples rather than textual explanations when solving new problems (Chi, Bassok, Lewis, Reimann & Glaser, 1989; LeFevre & Dixon, 1986; Pirolli & Anderson, 1985). Therefore, students believe in the power of worked-out examples for learning concepts and procedures.

Research evidence demonstrates that examples can be effective learning tools in a wide range of fields, including physics (Chi et al., 1989), algebra (Zhu and Simon, 1987), chemistry (Lee & Hutchison, 1998), computer programming (Pirolli,1991; Pirolli, & Anderson, 1985), and learning to use a word processing program (Catrambone, 1995). Most importantly for the present discussion, examples are effective for learning probability (Ross & Kennedy, 1990) and tests of significance (Quilici & Mayer, 1996). These studies have reported superior learning through use of examples compared to instruction without examples. In the case of mathematics, Sweller and Cooper (1985) found that students who studied worked examples both learned more quickly and performed significantly more accurately than students who learned by working problems. Whether students can generalize from examples and transfer the knowledge appropriately to novel problems, however, is clearly a more complicated question, with some studies finding positive transfer (e.g. Cooper & Sweller, 1987; Zhu & Simon, 1987), some finding none (e.g. Catrambone, 1995; Reed, Dempster & Ettinger 1985; Sweller & Cooper, 1985) and others finding transfer only under specific conditions (e.g. Atkinson, Catrambone, & Merrill, 2003; Chi et al., 1989; Ross & Kennedy, 1990).

The use of examples, therefore, can be a highly effective learning tool, but there are limitations and potential pitfalls in using examples. Sweller and Cooper (1985) found that students who had learned to solve algebra problems from examples solved new problems faster and with fewer errors than students who had not used examples when the new problems were similar in structure to the examples. When, however, the new problems differed in structure, even in superficial ways, the advantage for example-based learning disappeared. Reed et al. (1985) found that participants relied heavily on exact syntactic matches between example and problem and thus did not transfer learning to new problems of similar but not exact format, even when the experimenters provided a conceptual rationale for the example. Obviously, if students learn only a strict algorithm or recipe for solving structurally identical problems in which they can substitute values to get the right answer, there is no true understanding of the underlying concept. Thus, examples can be highly effective for developing generalized understanding, but the mere use of examples does not guarantee improved learning.

THE COGNITIVE BASIS OF LEARNING THROUGH EXAMPLES

What cognitive processes are involved in successfully learning from examples? The cognitive processes that likely occur when learning from an example appear in Figure 6.1. The diagram is plotted in two dimensions. The vertical dimension is abstractness to concreteness, with greater height indicating abstractness and lower height indicating concreteness. This dimension is almost certainly correlated with other dimensions that affect learning. A concept that is abstract to a learner is one that is unfamiliar, unclear, ambiguous or confusing, whereas a concept that is concrete is one that is familiar, clear and well-defined. The horizontal dimension is the time course of learning.

The diagram illustrates two common ways that teachers use examples. In the first, a novel, typically abstract, concept is explained to students followed by one or more examples. The examples are concrete, familiar, or simple illustrations of the concept. The students can then relate the abstract concept to the concrete examples. Finally, the students can use the examples to develop an abstract, conceptual understanding that they can apply to novel situations. In the second way, a teacher introduces examples to students without explaining the overall concept. The students must then arrive at an abstract conceptual understanding inductively from the examples.

The diagram demonstrates ways that examples should work, but it also indicates a number of ways that examples can fail as a learning tool. First, the examples may not be sufficiently concrete, familiar, or clear enough for students to understand. An example that is simple and

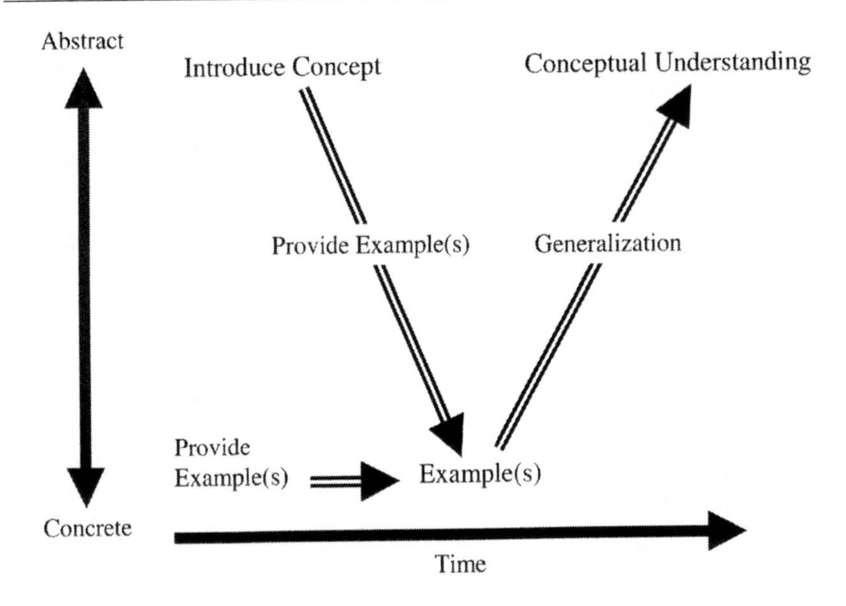

Figure 6.1. Learning from examples.

straightforward to the teacher may be opaque to the student. When teaching probability, I use the example of drawing a particular card from a deck, which has always been familiar to students. In recent years, however, due to the popularity of video games, I have encountered more students who are unfamiliar with a deck of cards, and my examples become useless.

A second way that examples can fail occurs when the students cannot relate the example to the concept or relate the example inaccurately to the concept. At a teaching conference once, I saw a presenter use a pile of sand on a table to demonstrate frequency distributions and skewness. Although no doubt engaging and memorable to students, the effectiveness of the example would depend on students being able to relate the sand to a sample or population of participants and the front of the table to values of a dependent variable.

A third potential problem with using examples occurs when students fail to generalize from the examples to a more conceptual understanding. Students will often confuse understanding the example with understanding the concept. Their understanding is bound to the context, story line, and other superficial aspects of the example (Ross, 1989b). They expect all subsequent problems to look like, and be solved like, the example. These students can neither recognize problems of the same class as the example that differs substantially in form, nor apply the principle effectively to solve the problem. Without a conceptual understanding, successful transfer of knowledge to new situations will not occur.

Assuming a teacher has avoided the pitfalls of using examples, what cognitive processes occur when learning from examples is successful? Analogical reasoning is at the heart of learning from examples (Chi et al., 1989; Zhu & Simon, 1987). First, learners must understand how the example is an instantiation of the concept they are trying to master. Second, learners must recognize the relation between novel problems and the examples that they have already learned. Analogical reasoning and problem solving is neither automatic nor spontaneous, but both effortful and situationally dependent (Chen & Mo, 2004; Gentner & Holyoak, 1997; Gick & Holyoak, 1980).

Cooper and Sweller (1987) demonstrated that studying worked examples is an effective learning strategy for novices for two reasons. First, examples are more conducive to the development of schematic knowledge than working problems. Working problems leads to highly effortful problem-solving strategies that are aimed at solving the problem rather than creating a problem-solving schema. A *problem-solving schema* is the organized knowledge needed to correctly categorize a problem and recall the steps needed for its solution. In other words, a student might solve a problem involving *t* tests without gaining any understanding of *t* tests. Studying a worked example, however, is much more conducive to learning when to use a *t* test and how to compute a *t* test. Second, studying worked examples promotes the development of automaticity in using the schematic knowledge. Automatic access and recall of schema facilitates generalization and transfer of knowledge. When teaching introductory statistics, the usual method of explaining a concept, working out an example or two, and then having students (try to) solve challenging problems may not be conducive to either schema formation or automation of recall (Sweller, van Merrienboer, & Paas, 1998).

THE PROPERTIES OF A GOOD EXAMPLE

Several lines of research address the issues of what properties constitute a good example and how examples should be used in the classroom. Bransford, Franks, Vye, and Sherwood (1989) stated that good examples should do two things: reveal the reasoning underlying the solution and highlight critical features of a concept. Critical features can be key aspects of a concept or key contrasts that differentiate between one concept and other similar concepts. Understanding examples requires a great deal of implicit knowledge about the rationale for taking specific actions, the actions themselves, and the outcomes of the actions. Examples and worked problems in textbooks routinely leave out rationale or explanation about why a problem is solved in a certain way (Chi et al., 1989). In class, teachers may provide examples but might not explain explicitly how the examples relate to a concept. Teachers might not highlight the reasoning and procedures used to classify and solve the example. Finally, they might fail to explain why one action is appropriate and others are not. If students lack the implicit knowledge and cannot construct the reasoning behind the example, then they can only follow the steps of the example mechanically (Atkinson et al., 2003; Chi et al., 1989; Reed and Bolstad, 1991; Reed et al., 1985).

There are practical reasons why full explanation and rationale are not usually included with examples. To be effective, examples are supposed to be simpler than the explanation of a concept. An example that contains exhaustive detail and rationale becomes too long and complex to understand (Ward & Sweller, 1990). Thus, examples must balance degree of explanation with complexity. Examples add to the amount of reading or listening students have to complete. The gain in learning from examples must justify the additional time required to try to understand them (Catrambone, 1995; Lee & Hutchison, 1998). The more detailed and complex the example, the longer it will take students to work through it.

In a series of studies, Ross (see Ross, 1989b for a summary), demonstrated the utility of breaking down an example into its surface and structural components. The structural aspects of an example are those features that determine the underlying principle needed to address the example. A change in the structural components changes the category of the example and the means used to solve the example. The surface or superficial aspects of an example can be changed without altering the category of the example or how the example is solved. Surface and superficial aspects of an example include the storyline, word choices, names of objects, or numerical values used. The literature on examples and analogies uses the terms *superficial components* and *surface components*, interchangeably. I prefer the term *surface components* because the influence of this aspect of an example on learning is hardly superficial. Table 6.1 contains an example about computing combinations and breaks it down into surface and structural components.

Ross (1989a, 1989b) demonstrated that even though the surface aspect of an example is irrelevant to classifying or solving an example, it is highly influential in what students learn from an example. When students study an example, they store a combination of the surface and structural aspects of the example in memory as one integrated memory. Students lack the expertise to separate the two components of the example. Teachers think of the example context as merely a means to illustrate a principle, but students see the context as part of the principle. Furthermore, the surface components are likely to be much more familiar and memorable to students.

Ross (1989b) demonstrated that the surface component of the example is much more effective in reminding students of an example at a later time than the structural component. Thus, if a statistics teacher used the example about electing officers in a gardening club in Table 6.1, the teacher would believe the example was about calculating combinations, but the students would believe the example is about calculating combi-

TABLE 6.1

Surface and Structural Components of Examples

A Teaching Example for Combinations

The local gardening club has 28 members. They need to elect two members to serve as officers. How many variations of different members can serve as officers?

Surface Components: A gardening club with 28 members electing two officers.

Structural Component: This is a combinations problem because it does not matter to which office a member is elected.

Problem 1

A pizza restaurant offers 12 different kinds of toppings. If you wanted to have a three topping pizza, how many variations could you have?

Surface Components: Different from the example.

Structural Components: Similar to the example.

Problem 2

The local gardening club has 28 members. They need to elect two members to serve as president and vice-president. How many variations of different members can serve as officers?

Surface Components: Same as the example.

Structural Components: Different from the example. Election of a president and vice-president makes this a permutations problem because it now matters to which office a member is elected.

nations when electing officers in a gardening club. At a later time, if the teacher wished to remind the students of the example, the teacher would be much better served by asking the students to remember the example about the gardening club than the example about combinations. On an exam, the teacher might give Problem 1 from Table 6.1, about combinations of toppings on a pizza, to test students' knowledge of combinations. To the teacher, the problem and example are similar because they are both about combinations, but to the student, the example and problem are quite different because one is about a gardening club and the other is about a pizza. The influence of the surface component explains the situation where a teacher provides a memorable example of a concept only to find later that the students remember the example, but not the concept it represents.

How might the surface context of an example help or hurt learning? Ross (1984) taught participants about probability using examples. He manipulated the storyline, part of the surface context, between the worked example and problem to be solved. Examples of different principles each had their own distinct story line. The story line of the problems could be appropriate (same as the relevant example), unrelated (a never before seen story line), or inappropriate (a story line used in an example of a different principle than the relevant one for the problem). He found huge differences in proportion of correct responses, with .77 for appropriate, .43 for unrelated and .22 for inappropriate. The results show that teachers should take the construction of the surface components of the examples they use seriously. Ross explained the results in terms of what he called *reminding*. The surface context was a potent reminding for participants of the relevant example. Both surface reminding and structural reminding occurred. Ross (1989b) reported that surface remindings were twice as common as structural remindings. Although surface remindings did improve problem solving, structural remindings, when they did occur, led to even better problem solving performance than surface remindings.

Matching storylines between examples and problems can strongly influence the identification of, or access to, a relevant principle for new problems, but the example must then be used to apply the principle to the novel problem. Ross (1989a, 1989b; Ross & Kennedy, 1990) demonstrated that reminding students of previous examples not only increased access to relevant principles, but it also improved the use and application of those principles in related problems. The surface context is not the only means to remind students of a relevant example, but it is a particularly potent one (Ross & Kennedy, 1990). Ross (1987, 1989b; Ross & Kennedy, 1990) proposed an analogical reasoning process of mapping the relevant example onto the problem, which leads to the development of general schematic knowledge and hence increases the likelihood of transfer. The problem reminds the student of the relevant example, and the act of reconstructing the example and applying it through analogy to the problem promotes the development of schematic, noncontextual knowledge. The key is to remind students which example is relevant when they try to solve a novel problem.

Statistics teachers can use the knowledge of how the surface context of a problem can cue the memory of a relevant example student learning to their advantage. Blessing and Ross (1996) have shown that some surface contexts are even indicative of type of problem. A story line about an urn and colored marbles almost certainly portends a probability problem. Nevertheless, one obvious shortcoming to using story lines to remind students of relevant examples for new problems is that students are likely to generalize their knowledge only to problems with the exact same story line.

A problem that shares a surface context with an example that is not appropriate for solving the problem can mislead and confuse students (Ross, 1984; 1989a). For example, in Table 6.1, the example uses the storyline of a gardening club electing officers. Say after working through the example, the teacher gives Problem 2 from Table 6.1 on the exam. A structural change has made this a permutations problem, even though the superficial story context of electing officers in a gardening club remains unchanged. Most students are likely to try to solve the problem as a combinations problem due to the surface context. Students would likely believe that the teacher was being underhanded and mean with that exam question.

Teachers can address the problem of confusing the surface and structural aspects of an example by presenting students with multiple examples of a concept that have varying surface contexts. Students would be forced to focus on common structural properties of the example rather than surface components. Quilici and Mayer (1996) demonstrated the effectiveness of using multiple, diverse examples in helping students learn to categorize correctly different kinds of statistical problems. The benefit of diverse examples was especially marked for weaker students. Stronger students appeared to focus on structural components spontaneously and needed only one example for effective learning. Likewise, Paas and van Merrienboer (1994) found that highly variable worked examples led to superior learning and transfer for solving geometry problems compared to low variability examples.

In an unpublished study, Leslie Nix, Marie Beeseley, and I conducted follow-up research on manipulating the surface context of examples to help students learn and generalize basic concepts of probability (Nix, Beesley, & Chew, 2002). We conducted the research using students in statistics classes offered by either psychology or mathematics. We explained four basic types of probability (marginal, joint, multiple outcome and conditional probabilities) and demonstrated how to calculate each one using an example. Participants next had to solve a set of problems. We manipulated the story contexts of both the examples and the first problem set. For some participants, the problems had the same story contexts as the relevant examples and for others the problems had completely different story contexts. Finally, participants tried to solve a set of transfer problems, which were probability problems with novel surface contexts. The results were consistent with the research already summarized. For the first problem set, participants solved problems best when the examples of each type of probability had a different story context (as opposed to all examples having a common story context) and when the problems used the same story context as the relevant example. In other words, story contexts both helped students to distinguish among the different kinds of probability and to access the relevant example when solving problems. The performance on the transfer problems, however, showed a different pattern of results. Transfer performance was best when the story contexts of the first set of problems were unrelated to the story contexts of the examples. The difference in story contexts between the examples and the first problem set seems to have forced students to process the structural components of the example and promoted generalization of knowledge. Our study confirmed that the surface context of examples, which are likely given little thought by teachers, can have a significant impact on initial learning and transfer.

Sweller and colleagues (Chandler & Sweller, 1991; Sweller, 1988; Sweller et al., 1998; Ward & Sweller, 1990) have amassed a large body of research demonstrating the impact of cognitive load on the effectiveness of examples. Cognitive load has two components: mental effort and mental load (Sweller et al., 1998). *Mental effort* refers to the amount of concentration, or conscious mental effort, that a student expends in trying to complete a task such as understanding an example. Mental effort is a limited resource in the human cognitive system. *Mental load* is a property of a problem or example and refers to the amount of concentration that a student has to expend in order to understand it. Mental effort is how much concentration a student is willing and able to focus on an example, whereas mental load is how much concentration on the part of the student a problem requires to be understood.

A central theme in Sweller's work is that if students have to expend virtually all of their limited cognitive resources in order to understand a problem or work an example, then there is little mental effort left in order to reflect on the example or problem and create general schematic knowledge. Thus, assigning students a series of challenging problems to solve a new mathematical concept may result in little to no learning because all of the students' mental effort was taken up in seeking a problem solution (Sweller & Cooper, 1985). Examples that impose a heavy cognitive load will not be effective (Chandler & Sweller, 1991). Sweller advocated a much greater emphasis on using well-designed, worked examples for teaching to reduce cognitive load and leave more mental resources available for generalizing knowledge and creating problem solving schema. The advantage for worked examples over problem solving is true only for novice learners. Once a certain level of expertise is achieved, solving problems can be as effective or more effective for learning (Kalyuga, Chandler, Tuovinen, & Sweller, 2001; Ward and Sweller, 1990).

Teachers must, therefore, design examples that minimize cognitive load—but how can teachers accomplish that goal? Sweller and colleagues distinguished between the intrinsic and extraneous cognitive load of examples (Sweller et al., 1998). *Intrinsic cognitive load* is mental effort required to understand the principle embodied by an example. *Intrinsic cognitive load* represents the minimum cognitive load needed to understand an example. *Extraneous cognitive load* is anything that adds to the overall cognitive load that is not essential to understanding the example. Extraneous cognitive load includes poorly designed instruction, activities that direct student attention away from the principle to be learned, and even the structure of the example itself. The instructor has control over extraneous cognitive load and should try to minimize it. For example, Sweller and colleagues have shown that examples that require students to integrate two sources of information, such as an illustration that is explained in the text of a book, increases cognitive load and diminishes the effectiveness of the example compared to examples where the text is integrated into the illustration (Chandler & Sweller, 1991). Furthermore, when there is redundant information in the text and illustration, the extraneous cognitive load increases and diminishes the example. A needlessly complex or obscure story line could contribute to extraneous cognitive load.

Intrinsic cognitive load, sometimes called germane cognitive load, is determined by what Sweller (Sweller, 1994; Sweller & Chandler, 1994) calls elemental interactivity. Material high in elemental interactivity has multiple components that must all be mastered simultaneously in order to understand the overall concept. Mathematics, and hence statistics, has high elemental interactivity because students must understand a large number of interrelated concepts. Material low in elemental interactivity is composed of elements that can be understood in isolation and learned serially. Learning new vocabulary words is an activity with low elemental interactivity. Intrinsic cognitive load is direct function of element interactivity. When elemental interactivity, and hence cognitive load, is low, then extraneous cognitive load probably will not matter. If elemental interactivity is high, which it is in statistics, then intrinsic cognitive load will be high, and extraneous cognitive load becomes critical to the effectiveness of examples.

Cognitive load theory has a number of practical implications for statistics instruction (Sweller et al., 1998). I have already discussed minimizing extraneous cognitive load by reducing redundancy and avoiding the splitting of attention. Sweller also recommended presenting goal-free problems for students to solve. Instead of presenting a problem and asking for a specific answer, ask the students to calculate the value of whatever variables they can. The open-ended question shifts the focus from problem solving to schema building. In order to ensure that students attend to the example, Sweller et al. suggested giving students partially worked examples to complete instead of completely worked examples.

USING EXAMPLES EFFECTIVELY

Once a teacher presents an example, how the students process it is of prime importance (Chi & Bassok, 1989; Chi et al., 1989). Chi et al. (1989) had participants study worked examples of physics problems and then solve new problems while thinking aloud. Chi et al. then analyzed the verbal protocols generated by the participants to determine the strategies the participants employed to use the examples in order to solve the problems, and how effective the strategies were (cf. Ericsson & Simon, 1993). Chi et al. divided students into "good" and "poor" groups based on their performance. Good students engaged in significantly more self-explanations when using the examples to solve the problems. Self-explanations occurred when students elaborated on an example to understand the underlying concept and actions involved in order to solve a related problem. Chi et al. found that good students spontaneously generated self-explanations, but the poor students did not. Good students not only had more self-explanations than weaker students, but also better self-explanations that specified missing information and rationale from the example. The self-explanations of poor students tended to be paraphrases of the example that did not address the underlying principle of the example. In Sweller's terms, self-explanations were related to the development of general problem-solving schema. In addition to more and better self-explanations, good students were also better than poor students at accurately monitoring their level of comprehension of the example. Although the work of Chi and colleagues makes a strong case for the relation between self-explanations and learning, the effectiveness of self-explanations may be dependent on the age and verbal abilities of the participants, and how easily verbalized the topic area is (Mwangi & Sweller, 1998).

Giving students examples is obviously not enough. For an example to be effective, students must elaborate upon it in a way that leads to deeper understanding. The work of Chi et al. (1989) raises the possibility that good examples might facilitate, invite, or even provoke self-explanations, at least for stronger students. If students do not spontaneously engage in self-explanations, how might teachers induce students to engage in reflection and elaboration of examples? There are several effective methods. One solution discussed earlier is to use multiple examples with varying surface contexts. This method forces students to focus on structural components, improving learning and transfer (Paas & van Merrienboer, 1994; Quilici and Mayer, 1996).

Catrambone (Atkinson et al., 2003; Catrambone, 1994, 1996) demonstrated the effectiveness of utilizing worked examples with solutions that are organized around conceptual modules or subgoals. Worked examples organized around subgoals led to superior transfer to examples with solutions not organized in this way. Furthermore, labeling the subgoals aids students in transferring the knowledge to novel problems, even if the labels themselves are meaningless. Subgoals that are oriented toward conceptual elaboration are superior to subgoals that emphasize computational steps.

Another way to induce students to generate self-explanations for examples is to use an orienting task that ensures that students will engage in elaborative and analogical thinking. Lee and Hutchinson (1998) examined this idea through the use of reflection questions that forced participants to focus on how the example was an instantiation of a concept. In a series of experiments, Lee and Hutchison examined three kinds of examples for learning how to balance chemical equations. *Case* examples simply gave the answers for an example, *Augmented* examples gave a partial rationale for the solution, and *Strategy* examples provided the complete rationale for examples. Furthermore, some participants answered reflection questions about the examples. Reflection questions were "why" questions about the example that forced participants to construct explanations for the examples. Both Augmented and Strategy examples led to better problem solving than Case examples. The Augmented examples, which contained only partial explanation of the example, were just as effective as Strategy examples that contained a complete rationale. Therefore, only part of the rationale was necessary for the example to be effective but some rationale led to better learning than no rationale at all. The use of reflection questions also led to superior performance and transfer to new problems. Consistent with prior research on examples, reflection questions were particularly helpful for students with weaker backgrounds in chemistry. Although Lee and Hutchison (1998) studied problem solving in chemistry, the results have implications for teaching statistics. I address those implications in an experiment described subsequently.

In summary, both teachers and students believe in the power of examples as a teaching tool, but their effectiveness, especially for developing general understanding, is not straightforward. The impact of an example depends on how the example (or series of examples) is constructed, how the student processes it, and the prior knowledge of the student. A good example minimizes extraneous cognitive load and contains at least part of the rationale for the solving the example. The student must elaborate on the example and understand the structural elements of the example to create general schematic knowledge. Examples are particularly useful for weaker students with little prior knowledge of a participant, and they have the greatest impact at the initial stages of learning (Catrambone, 1995).

THE EFFECT OF EXAMPLE TYPE AND REFLECTION QUESTIONS ON USING EXAMPLES FOR LEARNING STATISTICS

In an unpublished study, Darcy Dunbar, Jenny Leach and I tested the effectiveness of using reflection questions in combination with different kinds of examples for learning statistics (Leach, Dunbar & Chew, 2000). We built on the work of Lee and Hutchinson (1998), who studied examples and reflection questions for learning chemistry. We used three statistical tests of significance as our content: one sample *z* test, one sample *t* test, and *t* test with independent samples. In learning about these tests, students often confuse the conditions under which each test is used as well as the procedures.

There were two independent variables crossed in a 2 × 2 factorial design. The kind of example could be a Case example or a Strategy example. The example was followed by either reflection questions (Reflection condition) or control questions (Control condition). Case examples simply solved the problem without any rationale or explanation. Strategy examples provided a rationale for why that particular test was appropriate. Furthermore, the Strategy example emphasized the key differences among the three tests; specifically, when each is used and how one differs from the others. In the Reflection condition, participants answered questions that forced the students to think through and elaborate on the structural elements of the example, such as "Why did we choose to use a one-sample z test for this problem?" In the control condition, participants answered questions about the example that did not require elaboration, such as "Is this a real experiment or a quasi-experiment?"

We made three predictions. First, consistent with Lee and Hutchison (1998), we predicted that Strategy examples should lead to better performance than case examples. Second, we predicted that Reflection questions should lead to better learning than Control questions. Third, we predicted that the combination of Strategy examples and Reflection questions should lead to the highest overall performance.

A sample of 21 psychology majors, all enrolled in an upper level required statistics class for psychology, first read a textual explanation of each statistical test accompanied by the assigned example and question. The worked example for each of the three statistical tests used a different story context, but the contexts were kept constant across groups. After studying the text and examples and answering the questions for all three types of tests, participants attempted to solve four problems. There was at least one of each kind of kind of statistical test.

The teacher of the class graded the performance on the four problems as if grading an exam question. Each problem was worth up to 10 points for a correct and complete answer. The teacher was blind to the identity or condition of the participants.

The mean score correct for each group appears in Table 6.2. There was a significant interaction, $F(1,17) = 5.82, p < .05$. As predicted, the combination of Strategy examples and Reflection questions led to the highest performance. Surprisingly, the next best performance was from using Case examples and Control questions. The participants who had Strategy examples with control questions or Case examples with reflection questions performed the poorest with no significant difference between the two conditions.

Participants presented with the combination of elaborated Strategy examples and reflection questions solved the greatest number of problems. These results are consistent with the results of Lee and Hutchinson (1998) in showing that both the type of example and the way the example is processed influence performance. What is striking in our results is

TABLE 6.2

Mean Problems Solved by Example Type and Reflection Question

Example Type/Reflection Question	*Mean Score*
Case Examples/Reflection Questions	11.67
Case Examples/Control Questions	16.40
Strategy Examples/Reflection Questions	22.8
Strategy Examples/Control Questions	12.2

that a mismatch between kind of example and kind of question led to the poorest performance. Participants given an elaborated Strategy example but not required to process the rationale performed relatively poorly, as did participants given an unelaborated Case example but required to try to provide a rationale for it. Participants given the unelaborated examples and allowed to elaborate on them on their own performed better than the two mismatched groups.

The results demonstrate the importance of matching the right kind of example with the right kind of encoding process. Based on these results, giving participants elaborated examples without having them reflect on them actually hinders performance. Strategy examples possess a higher cognitive load than Case examples because they contain the additional rationale and explanation. The higher cognitive load may deter some students from spontaneously generating self-explanations without guidance from the teacher. On the other hand, giving students a minimal example and then directing them to construct rationales for the examples using reflection questions also leads to poor performance. The poor performance might be due to student frustration from being asked to answer a question that the example does not explicitly address. A second possibility is that students began to construct a rationale for the example when it was given, but the reflection question somehow interfered with the process by distracting students from their own explanations or confusing them about the best way to elaborate on the example. A practical implication of these results for statistics teachers is that it is incorrect to think of one example being better than another or one way of processing an example being better than another. Teachers must consider the interaction between the kind of example and how students process it.

A MODEL FOR DESIGNING EXAMPLES

To promote best practices for the use of examples, I propose the model of example design shown in Figure 6.2. The model is based on the research I have reviewed in this chapter and includes all the factors that influence the effectiveness of examples. In designing and using examples, teachers must consider how engaging an example is for students. Teachers must assess the cognitive load an example imposes on the students and try to minimize extraneous cognitive load. The model shows that one way of achieving both high engagement and low cognitive load is to use examples familiar to students, such as using a deck of cards for probability examples. Teachers must make sure that students engage in elaboration and self-explanation of the example. Stronger students might do this spontaneously, but weaker students may need guidance. Teachers must balance the amount of rationale included in an example with the cognitive load of an example and how the students will process the examples. Teachers must also design examples that facilitate transfer and application. They can use multiple examples with varying surface contexts or

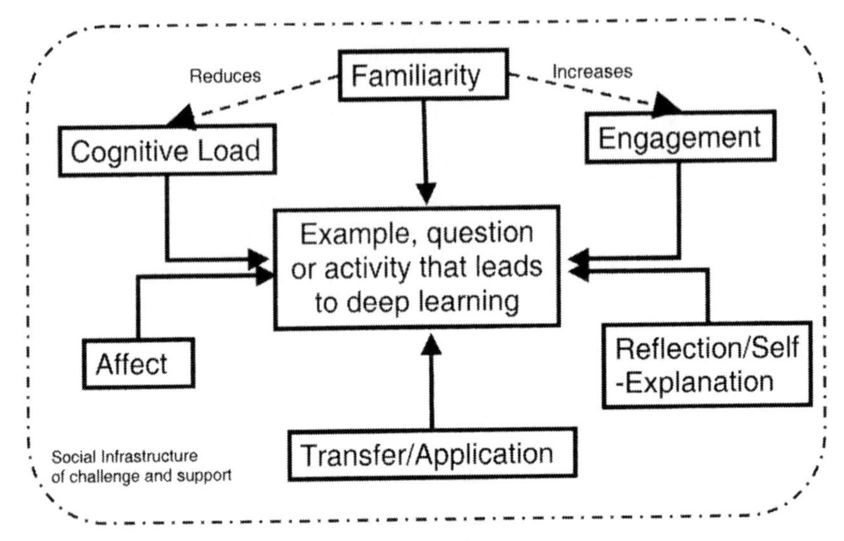

Figure 6.2. A model for designing examples.

examples broken down into subgoals. The model includes two factors that I have not discussed in this chapter, but are likely important factors. The affective component of an example likely influences an example's memorability and later accessibility. The classroom atmosphere must also enable and encourage the elaborative processing of examples and problems. Finally, the factors in the model do not occur in isolation but in interaction with each other. The literature is clear that the best kind of example depends on the combined action of many factors, such as the prior knowledge of the student, the kind of example, and the kind of processing in which the student engages. Teachers must take all these factors together to design effective examples for teaching statistics.

The research reviewed in this chapter shows that teachers cannot create examples haphazardly and assume that one example is as good as another or assume that the way the example is presented to or processed by the student does not matter. Teachers must design and implement examples to achieve a particular kind of learning on the part of the student. Ideally, a teacher can design a series of examples that will scaffold learning from the initial stages to generalized understanding. It is time to move beyond the intuitive use of examples to the implementation of best practices for the design and use of examples to achieve deep understanding.

ACKNOWLEDGMENTS

The work presented here was made possible by grants from the Carnegie Foundation for the Advancement of Teaching, the Pew Charitable Trusts, and Samford University. In particular, I thank Lee Shulman, Pat

Hutchings, the Pew Few, and John Harris for making this research possible. I also thank Jenny Leach, Leslie Nix, Darcy Dunbar, Marie Beesley, Gareth Dutton, and all my statistics students for collaborating on the research. Finally, I thank Dana Dunn, Barney Beins, and Randy Smith for their patience and thoughtful comments on earlier versions of this work.

Correspondence concerning this chapter should be addressed to Stephen L. Chew, Department of Psychology, Samford University, 800 Lakeshore Drive, Birmingham, Al 35229-2308 or by e-mail at slchew@samford.edu

REFERENCES

Atkinson, R. K., Catrambone, R., & Merrill, M. M. (2003). Aiding transfer in statistics: Examining the use of conceptually oriented equations and elaborations during subgoal learning. *Journal of Educational Psychology, 95*, 762–773.

Blessing, S. B., & Ross, B. H. (1996). Content effects in problem categorization and problem solving. *Journal of Experimental Psychology: Learning, Memory, and Cognition, 22*, 792–810.

Bransford, J. D., Franks, J. J., Vye, N. J., & Sherwood, R. D. (1989). New approaches to instruction: Because wisdom can't be told. In S. Vosniadou & A. Ortony (Eds.), *Similarity and analogical reasoning* (pp. 470–497). New York: Cambridge University Press.

Catrambone, R. (1994). Improving examples to improve transfer to novel problems. *Memory and Cognition, 22*, 606–615.

Catrambone, R. (1995). Following instructions: Effects of principles and examples. *Journal of Experimental Psychology: Applied*, 1, 227–244.

Catrambone, R. (1996). Generalizing solution procedures learned from examples. *Journal of Experimental Psychology: Learning, Memory, and Cognition, 22*, 1020–1031.

Chandler, P., & Sweller, J. (1991). Cognitive load theory and the format of instruction. *Cognition and Instruction, 8*, 293–332.

Chen, Z., & Mo, L. (2004). Schema induction in problem solving: A multidimensional analysis. *Journal of Experimental Psychology: Learning, Memory and Cognition, 30*, 583–600.

Chi, M. T. H., & Bassok, M. (1989). Learning from examples via self-explanations. In L. B. Resnick (Ed.), *Knowing, learning, and instruction: Essays in honor of Robert Glaser* (pp. 251–282). Hillsdale NJ: Lawrence Erlbaum Associates.

Chi, M. T. H., Bassok, M., Lewis. M. W., Reimann, P., & Glaser, R. (1989). Self-explanation: How students study and use examples in learning to solve problems. *Cognitive Science, 13*, 145–182.

Cooper G., & Sweller, J. (1987). Effects of schema acquisition and rule automaton on mathematical problem-solving transfer. *Journal of Educational Psychology, 79*, 347–362.

Ericsson, K. A., & Simon, H. A. (1993). *Protocol analysis: Verbal reports as data* (rev. ed.). Cambridge, MA: Bradford Books/MIT Press.

Gentner, D., & Holyoak, K. J. (1997). Reasoning and learning by analogy: Introduction. *American Psychologist, 52*, 32–34.

Gick, M. L., & Holyoak, K. J. (1983). Analogical problem solving. *Cognitive Psychology, 12*, 306–355.

Gick, M. L., & Holyoak, K. J. (1983). Schema induction and analogical transfer. *Cognitive Psychology, 15*, 1–38.

Kalyuga, S., Chandler, P., Tuovinen, J., & Sweller, J. (2001). When problem solving is superior to studying worked examples. *Journal of Educational Psychology, 93*, 579–588.

Leach, J. A., Dunbar, D. M., & Chew, S. L. (2000). *The effect of different types of examples and reflection questions on learning*. Unpublished manuscript, Samford University, Birmingham, Alabama.

Lee, A. Y., & Hutchison, L. (1998). Improving learning from examples through reflection. *Journal of Experimental Psychology: Applied, 4*, 187–210.

LeFevre, J., & Dixon, P. (1986). Do written instructions need examples? *Cognition and Instruction, 3*, 1–30.

Mwangi, W., & Sweller, J. (1998). Learning to solve compare word problems: The effect of example format and generating self-explanations. *Cognition and Instruction, 16*, 173–199.

Nix, L. M., Beesley, J. M., & Chew, S. L. (2002). *Effects of superficial context on the effectiveness of teaching examples*. Unpublished manuscript, Samford University, Birmingham, Alabama.

Paas, F. G. W. C., & van Merrienboer, J. J. G. (1994). Variability of worked examples and transfer of geometrical problem-solving skills: A cognitive-load approach. *Journal of Educational Psychology, 86*, 122–133.

Pirolli, P. (1991). Effects of examples and their explanations in a lesson on recursion: A production system analysis. *Cognition & Instruction, 8*, 207–259.

Pirolli, P. L., & Anderson, J. R. (1985). The role of learning form examples in the acquisition of recursive programming skills. *Canadian Journal of Psychology, 32*, 240–272.

Quilici, J. L., & Mayer, R. E. (1996). Role of examples in how student learn to categorize statistics word problems. *Journal of Educational Psychology, 88*, 144–161.

Reed, S. K., & Bolstad, C. A. (1991). Use of examples and procedures in problem solving. *Journal of Experimental Psychology: Learning, Memory and Cognition, 17*, 753–766.

Reed, S. K., Dempster, A., & Ettinger, M. (1985). Usefulness of analogous solutions for solving algebra word problems. *Journal of Experimental Psychology: Learning, Memory and Cognition, 11*, 106–125.

Ross, B. H. (1984). Remindings and their effects in learning a cognitive skill. *Cognitive Psychology, 16*, 371–416.

Ross, B. H. (1987). This is like that: The use of earlier problems and the separation of similarity effects. *Journal of Experimental Psychology: Learning, Memory and Cognition, 13*, 629–639.

Ross, B. H. (1989a). Distinguishing types of superficial similarities: Different effects on the access and use of earlier problems. *Journal of Experimental Psychology: Learning, Memory and Cognition*, 15, 456–468.

Ross, B. H. (1989b). Remindings in learning and instruction. In S. Vosnidou & A. Ortony (Eds.), *Similarity and analogical reasoning* (pp. 438–469). New York: Cambridge University Press.

Ross, B. H, & Kennedy, P. T. (1990). Generalizing form the use of earlier examples in problem solving. *Journal of Experimental Psychology: Learning, Memory and Cognition, 16*, 42–55.

Sweller, J. (1988). Cognitive load during problem solving: Effects on learning. *Cognitive Science, 12*, 257–285.

Sweller, J. (1994). Cognitive load theory, learning difficulty, and instructional design. *Learning & Instruction, 4,* 295–312.

Sweller, J., & Chandler, P. (1994). Why some material is difficult to learn. *Cognition and Instruction,* 12, 185–233.

Sweller, J. & Cooper, G. A. (1985). The use of worked examples as a substitute for problem solving in learning algebra. *Cognition and Instruction, 2,* 59–89.

Sweller, J., van Merrienboer, J. J. G., & Paas, F. G. W. C. (1998). Cognitive architecture and instructional design. *Educational Psychology Review, 10,* 251–296.

Ward, M., & Sweller, J. (1990). Structuring effective worked examples. *Cognition and Instruction, 7,* 1–39.

Zhu, X., & Simon, H. A. (1987). Learning mathematics from examples and by doing. *Cognition and Instruction, 4,* 137–166.

Chapter 7

Designing an Online Introductory Statistics Course

Charles M. Harris, James G. Mazoué,
Hasan Hamdan, and Arlene R. Casiple
James Madison University

College students of the Net Generation embrace technology as a viable means for achieving their goals of experiential learning and social interaction through person-to-person communication and dialogue with faculty (Howe & Strauss, 2000). During the past decade, interaction between faculty and students within online computer-mediated instruction emerged in two distinct formats (Shank & Sitze, 2004; Velsmid, 1997). One format, *asynchronous instruction*, is place and time independent. At different times and from different places, instructors and students post information and participate in electronic dialogues. The other format, *synchronous instruction*, is place independent but time dependent. Although they are in different places, instructors and students interact concurrently at specified times. Kuh's (1994) description of the Net Generation as eager to experience learning in a dynamic social setting underscores the importance of expanding the opportunities for students to actively engage peers and their instructors.

The James Madison University (JMU) model for distance education combines both asynchronous and synchronous interaction to provide online instruction that incorporates the traditionally valued structure and functions typical of conventional, face-to-face instruction. The model integrates an expanded set of Web-based technologies and principles of sound pedagogy within a defined theoretical framework. This chapter describes the application of the JMU model to the conceptualization and construction of an introductory statistics course. However, the model is applicable for a variety of online courses: introductory and advanced, concept-oriented and skill-oriented, lecture and lab-based courses (Harris, 2005).

The two-stage JMU model consists of conceptualizing online courses and constructing online courses. The theoretical framework, expanded set of technologies, and issues related to accessibility constitute the discussion of conceptualizing online courses. The discussion of constructing online courses focuses on integrating five Web-based technologies with Chickering and Gamson's (1987) seven principles for good practice in undergraduate education.

CONCEPTUALIZING AN ONLINE INTRODUCTORY STATISTICS COURSE

Theoretical Framework

Whether in online or conventional, face-to-face settings, theoretical conceptualizations of the nature of teaching and learning are central to the instructional process (Holmberg, 1989). *Information processing theory* and *Piagetian-type constructivism* exemplify a dualistic conceptualization of online instruction. Information processing theory is based on the objectivist view that knowledge is external to the one who is doing the learning. Therefore, the goal of instruction is to adequately represent the meaning and structure of a specified domain of knowledge in order to facilitate a student's accurate acquisition of that knowledge (Duffy & Jonassen, 1992). Applied to online instruction, information processing theory might take the form of extensive notation and referencing to domains of knowledge. The assumption underlying information processing theory is that exposure to knowledge is the necessary and sufficient condition for acquiring knowledge. Ayersman's (1995) description of online learning as associative, nonlinear, and hierarchical in structure exemplifies the information processing approach.

In contrast, Piagetian constructivism conceptualizes the structure of one's knowledge and learning as self-regulated adaptation of information to internal networks of knowledge previously acquired by an individual (Jonassen, 1999). Whereas information processing theory conceptualizes learning as the assimilation of externally structured information, Piaget postulated systematic cognitive processes as the necessary and sufficient mechanisms by which individuals learn, that is, internalize information and assign meaning to their experience. Applied to online instruction, Piagetian constructivism might take the form of presenting questions, problems, and tasks for self-paced consideration by individuals who assign meaning, structure, and value to the outcomes of their efforts.

Our view is that students and instructors working online are not served well by the external–internal dichotomy of information processing versus Piagetian-type constructivism theories. Vygotsky's (1962) theory of social constructivism is a more interactive and dynamic model for online instruction. Key concepts in Vygotsky's theory, also valued by college students of the Net Generation, include the role of modeling, experiential learning, and social interaction. Vygotsky described the nature of knowledge in terms of spontaneous and scientific concepts, analogous to the novice-to-expert conceptualization of learning (Bruning, Schraw, & Ronning, 1999). *Scaffolding* is Vygotsky's term for the systematic, incremental process by which students progress from novice to expert levels of comprehension of specified domains of knowledge. Scaffolding results from social interaction between students and instructors. Thus, learning is an internal, incremental, cognitive process that is facilitated by interaction with sophisticated others.

Applied to online learning, a Vygotskian approach would initially present a comprehensive and explicit description of a task or assignment followed by a demonstration or illustration of the expected outcome. Subsequently, students would replicate solving or performing the assigned task. At this point, an information processing approach might consider a student's response to be indicative of sufficient learning. However, Vygotsky required a further demonstration of learning. The student must now provide a sufficiently accurate narrative of learning. That is, in addition to stating or demonstrating an expected outcome, students must describe how the task was achieved and explain why the response was relevant to the assigned task.

At this point, one might comment that the previous depiction of Vygotsky's social constructivism should ideally characterize all instruction. We agree that it should; however, we do not agree that it does. Technology-based online instruction can easily be reduced to mere dissemination of information (Duffy & Jonassen, 1992). In contrast to mere dissemination and aligned with the values of college students of the Net Generation, the theoretical framework of the JMU model facilitates dynamic, interactive online learning experiences that lead to higher order cognitive processing of the specified domain of knowledge.

Instructional Technologies

When designing a pedagogically sound online course, the initial task is selection of appropriate course management software (Elbaum, McIntyre, & Smith, 2002). One may choose from a variety of course management systems, such as Angel 6.2, Blackboard 6.2, Moodle 1.4, TopClass, or WebCT. We recommend that readers view the respective Web sites for detailed descriptions of distinctive features of these portal systems. (Consult the CD accompanying this volume for contact information and a comparison of the course management systems.)

After selecting a course management system, the next step is selection of ancillary technologies that will enable the creation of online courses with the pedagogical structure and functions that are typical of conventional, face-to-face courses. The aggregate functions of a pedagogically sound online course include (a) information dissemination with links to any Web-based source, (b) live audio conferencing for purposes of group instruction and individual mentoring, (c) video lectures that approximate face-to-face lectures, (d) small-group experiential learning, (e) writing assignments, and (f) performance assessment.

Selection of technologies for conducting online courses tends to be based on pragmatic considerations (Shank & Sitze, 2004); that is, decisions tend to be influenced by an instructor's familiarity with specific software or on the availability of technical support. In contrast, the JMU model utilizes five technologies that enable course design and construction based on relevant theory and sound pedagogy: Blackboard, Centra Symposium, Tegrity Web Learner, Respondus, and Adobe Acrobat.

Blackboard 6.2. The Blackboard Learning System is a Web-based learning management system that enables dynamic course management, an open architecture for customization and interoperability, and a scalable design that allows for integration with student information systems. Functions include (a) information dissemination, (b) scalable e-mail, (c) discussion boards, (d) small-group activities, (e) file sharing and transmission of written reports, and (f) performance assessment with instant feedback for students and summary statistics for instructors.

Centra Symposium 6.1. Centra supports live audio sessions, using voice-over IP, that accelerate knowledge transfer simultaneously for individuals and groups through collaborative learning. A Centra session may be recorded for subsequent replay by instructors and students. Functions include (a) live audio conferencing for purposes of group instruction and individual mentoring, (b) text-based instant messaging, and (c) information dissemination with links to any Web-based source.

Tegrity Web Learner. Tegrity Web Learner is a multimedia content authoring system that creates streaming media delivered over the Internet. A distinctive feature of the Web Learner is synchronized video lectures with simultaneous displays of the instructor and PowerPoint-based text with ancillary linkages. Tegrity's bandwith efficiency shows for 28K, 56K, and 100K viewing with a 28K audio-only option.

Respondus 2.0. Respondus is a Windows application that enables the offline creation and management of quizzes and tests. With similarities to a word processor, the Respondus Editor includes a spell checker, table and equation editors, and full media support. Respondus can import text from a word processor and seamlessly publish directly to Blackboard.

Adobe Acrobat 6. Acrobat enables the conversion of word processor-generated or scanned documents into portable document files (PDF) that preserve the format and style of original documents. PDFs can be secured and read by anyone, regardless of hardware and software platforms.

Web Accessibility Guidelines

Designers of online learning experiences are responsible for insuring equal access for students of all abilities. Updates about applications of the Americans with Disabilities Act, www.usdoj.gov/crt/ada, and its oversight by the federal Office of Civil Rights, www.hhs.gov/ocr, are accessible through their respective Web sites. Comprehensive guidelines for accessing the content of online applications are under development by the Web Accessibility Initiative of the World Wide Web Consortium, www.w3.org. We recommend that designers of online courses consult these resources in the early stages of course development. The challenge of integrating multiple technologies and sound pedagogy within online learning is made more complex by the diverse needs of students. In view of universal access to Web content, designers of online learning must ensure that their courses accommodate students with visual, auditory, motor, and cognitive impairments.

CONSTRUCTING AN ONLINE INTRODUCTORY STATISTICS COURSE

Equivalency with the processes and outcomes typical of conventional, face-to-face courses should be the guiding principle throughout the construction of online courses. Technology will have an essential, instrumental role in the construction of an online course; however, the selection and use of technology should be driven by principles of sound pedagogy. To that end, the JMU model incorporated the seven principles for good practice in undergraduate education developed under the leadership of Chickering and Gamson (1999). The seven principles for good pedagogy are applicable to discipline-based, multidisciplinary, and interdisciplinary courses offered in both conventional and online settings. In the discussion that follows, the content and examples are representative of an online introductory statistics course. The strategies and techniques exemplify numerous ways to achieve the dynamic interaction that is central to Vygotsky's (1962) social constructivism and valued by college students of the Net Generation.

Principle 1: Encouraging Student–Faculty Interaction

Interaction between students and their instructors is an essential element of the processes we label *teaching and learning* (Holmberg, 1989). Facilitating student-to-instructor and student-to-student interaction in an online introductory statistics course is especially important because of the aversion to statistics observed in many and the anxiety observed in some (Buxton, 1991; Hackworth, 1985; Zaslavsky, 1994). The JMU model features interaction in online courses that is equivalent with conventional instruction by incorporating several technologies that support both asynchronous and synchronous interaction.

The decision to adopt Blackboard as the course management system within the JMU model was influenced by the multiple functions supporting asynchronous interaction. The Blackboard home page appears in Figure 7.1. Several Blackboard functions support instructor-to-student and student-to-student text-based interaction. Students access text-based information conveyed by instructors, in secure PDF format, through the announcements, faculty information, course information, and course documents functions within Blackboard. The assignments, discussion board, and communication functions enable text-based asynchronous exchanges and discussions by students and instructors. The Blackboard Communication page appears in Figure 7.2. The option to e-mail all students simultaneously or a select subset of students is a prime function of the communication page, increasing the frequency of interaction and substantive exchanges.

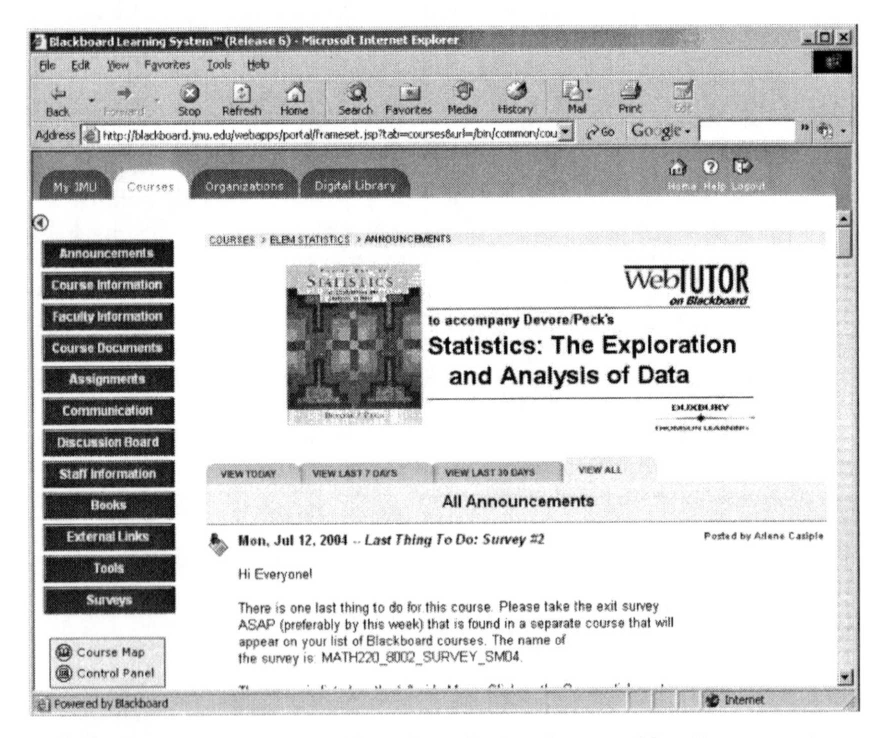

Figure 7.1. The Blackboard Home Page depicts the array of functions comprising the course management system.

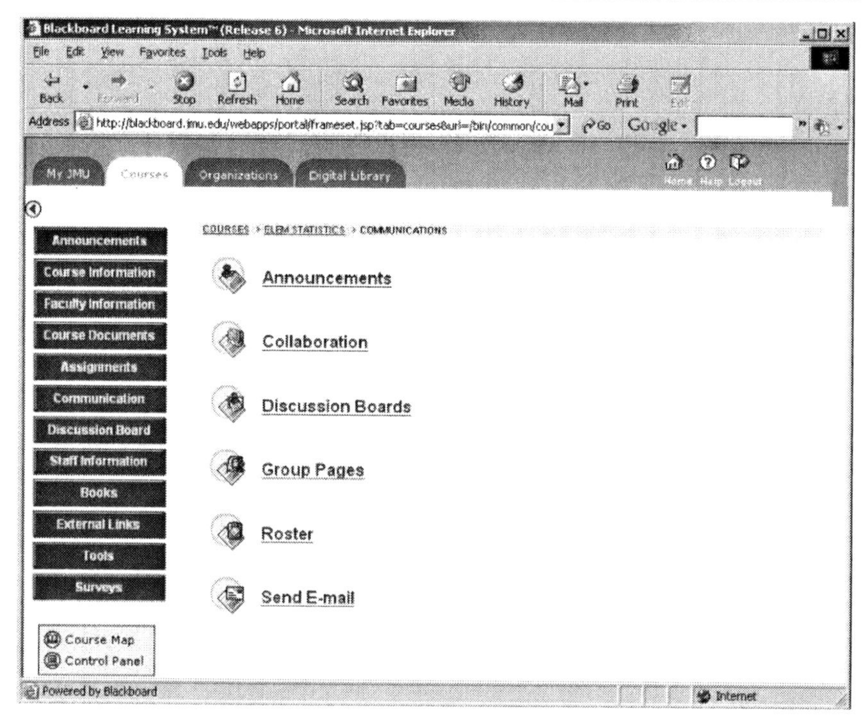

Figure 7.2. The Blackboard Communications Page depicts a variety of options for asynchronous, text-based communication among students and instructors

Tegrity Web Learner supports asynchronous interaction. Tegrity enables the delivery of brief streaming video lectures that feature synchronized displays of instructors delivering lectures and the corresponding PowerPoint-based text with ancillary linkages. A typical Tegrity slide appears in Figure 7.3. Putting a human face on an otherwise text-based experience is a distinctive feature of the Tegrity system. Using the control panel and slide bar, students have the following options: (a) pausing the lecture for taking notes, (b) reversing the lecture to replay a specific section, and (c) fast forwarding the lecture. Tegrity's bandwith efficiency includes viewing options at 28K and 100K. Tegrity software is a free download for students.

Synchronous, verbal interaction is a function of Centra Symposium (see Fig. 7.4). At specified times, instructors verbally engage their online students for purposes of minilectures, question-and-answer sessions, and office hours. Adding a human voice to an otherwise text-based experience is a distinctive feature of the Centra system. After convening a Centra session, instructors have the following options: (a) verbally addressing all students, (b) allocating a microphone to a specific student, (c) verbally responding to the question or comment of a specific student,

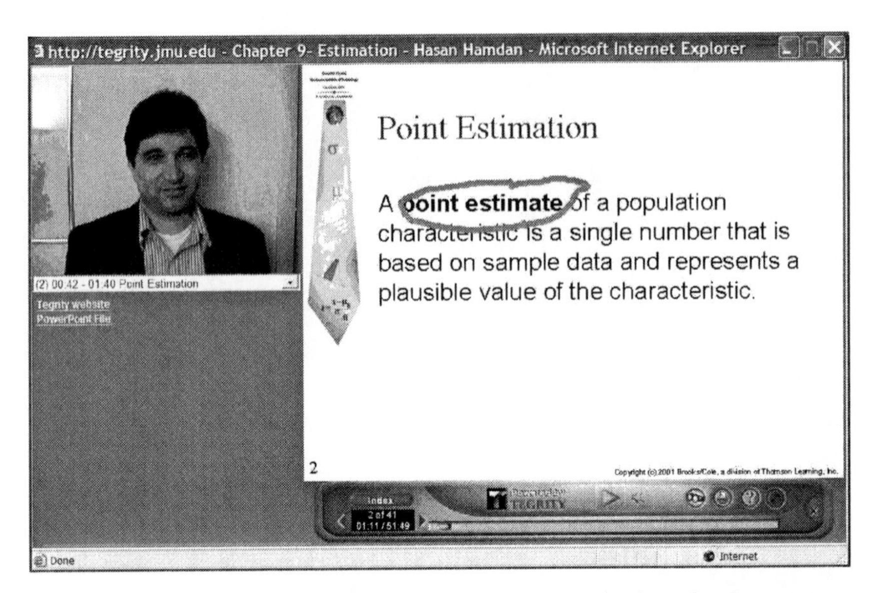

Figure 7.3. A typical Tegrity slide depicts synchronized video of an instructor and display of PowerPoint-based text.

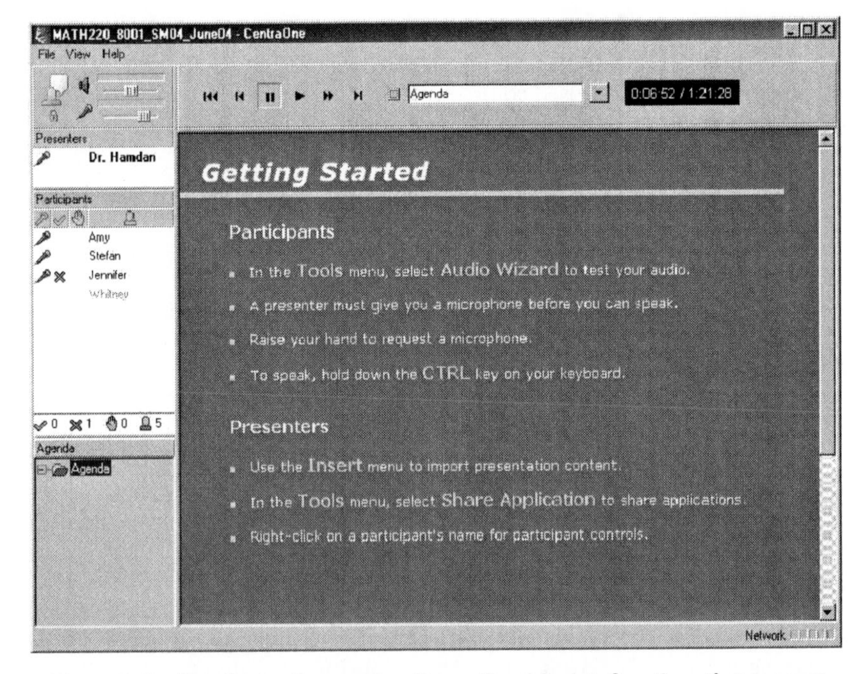

Figure 7.4. The Centra Symposium Home Page depicts functions that support synchronous, live audio sessions using voice-over IP.

(d) organizing students into small virtual discussion groups, (e) monitoring the dialogue within each discussion group, and (f) verbally responding to the question or comment of a specific discussion group. Centra sessions can be recorded and archived for subsequent review of the concepts discussed in the session and for viewing by students who were absent from that session. Centra software is a free download for students.

Principle 2: Encouraging Cooperation Among Students

Cooperative learning experiences that incorporate group rewards and individual accountability are effective strategies for improving motivation to learn and for achieving instructional objectives (Slavin, 1995). Strategies for implementing online cooperative learning include STAD (Student–Team Achievement Divisions), the jigsaw classroom (Aronson & Patnoe, 1997), learning together, group investigation, and cooperative scripting (Santrock, 2001). In addition to being consistent with learning in Vygotsky's social constructivism, cooperative learning activities, in an online introductory statistics course, support a variety of learning outcomes. An interactive online learning environment compensates for the lack of face-to-face interaction typical of conventional classroom settings.

Additionally, organizing students into small collaborative groups increases dialogue among students about complex statistical concepts. As members of a small group, students demonstrate a greater willingness to generate questions. Diverse abilities within a group facilitate assigning complex statistical problems. And, a more general outcome is improved communication skills as students develop increased clarity in expressing their questions and their comments about complex statistical concepts. The Group Pages option on the Communication page (see Fig. 7.2) in Blackboard supports asynchronous small group, cooperative learning experiences. Within each small group, members communicate by means of e-mail, discussion board and file exchange functions that are unique to their group.

Group projects that offer a variety of learning outcomes within an online introductory statistics course include (a) describing the distribution of qualitative and quantitative variables, (b) illustrating the concept of *probability*, (c) calculating correlation coefficients and the regression equation, and (d) drawing inferences about the parameters of a population. (Consult the CD accompanying this volume for examples of group projects suitable for online courses.) Brief descriptions of three cooperative learning assignments follow.

Summarize the Numerical and Graphical Properties of a Univariate Data Set. Numericalsummaries must include calculating measures of central tendency (mean, median, and mode) and measures of variability (range, standard deviation, variance, and interquartile range). Graphical summaries must include constructing box plots, stem-and-leaf displays, dot plots, and bar graphs.

Analyze aBivariate Data Set. Calculate the correlation coefficient, find the regression equation, and calculate the standard error of estimate.

Apply the Central Limit Theorem. Generate data from different distributions and investigate the shape and variability as the sample size changes.

Centra (see Fig. 7.4) supports synchronous cooperative online learning experiences among students in geographically different locations. Instructors can organize students participating in a particular Centra session into small virtual groups to work on specific tasks. By monitoring the dialogue within the respective groups, instructors can identify questions and problems common among students in the various groups. Then, a virtual microphone can be assigned to each virtual group to facilitate a collaborative clarification of the question and cooperative development of the solution.

Principle 3: Encouraging Active Learning

We recommend that the facilitation of active learning in an online introductory statistics course include efforts to minimize the effect of two potential hindrances. First, distance learning environments tend to be perceived as impersonal and individually isolating (Shank & Sitze, 2004). Also, a general aversion to mathematical and statistical concepts tends to subvert individual student initiative for learning (Zaslavsky, 1994). These challenges can be minimized when a high level of interaction among students and instructors is complemented by clearly stated learning objectives, well-designed assignments, and frequent opportunities for students to select self-paced practice exercises.

A variety of functions within Blackboard support active learning through self-paced problems and practice exercises constructed by the instructor or selected from a standard textbook. For example, instructors facilitate active learning by posting practice exercises under the Assignments function; or instructors post practice quizzes with answer keys under Course Documents. We recommend using Respondus for offline construction of tests and quizzes. Blackboard allows a variety of options when constructing a test, for example, setting a 60 min time limit for a 40-item test. Instructors have the option of making all items visible for the duration of a test or restricting viewing to individual items with no option to return to an item once a response has been entered. Post explanations for correct and incorrect responses to test items to encourage students to practice self-monitoring and further review of the concepts being assessed. Upon completion of a test, Blackboard immediately posts a student's score in My Grades under Tools (see Fig. 7.1).

Principle 4: Giving Prompt Feedback

The Net Generation values experiential learning that features a rapid pace and immediate responses (see Barron et al., chap. 10, this volume; Howe & Strauss, 2000). The Pew Internet & American Life Project (Jones, 2002) reported one in five of today's college students began using computers between the ages of 5 and 8, and all of them used computers by the time they were 16 to 18 years of age. Nationwide, approximately 90% of college students have gone online, compared with about 60% of the general population. With instant messaging, chat rooms, e-mail, and streaming video through cell phones and computers, feedback is a common life experience for college students of the Net Generation.

Just-in-Time Teaching (JiTT) is a Web-based teaching and learning strategy that facilitates timely interaction between students and instructors (Benedict & Anderton, 2004; refer also to Stoloff et al., chap. , this volume). Students respond electronically to Web-based assignments that are due shortly before the next class. The instructor reads students' submissions just in time to adapt instruction to facilitate student learning. The JiTT feedback model is an application of Chickering and Gamson's (1999) seven principles for good practice in undergraduate education. Apply JiTT within an online course by posting under Assignments, in Blackboard (see Fig. 7.1), questions or problems designed to promote active learning. Within the Assignments function is an Assignment option that requires students to enter a response in order to exit the activity. Students post their responses by a predetermined time prior to the beginning of the next scheduled Centra (see Fig. 7.4) session. Instructors review the students' responses to identify areas of understanding and misunderstanding. The just-in-time feedback enables instructors to adapt the focus of the live Centra discussion to actively engage students at their respective levels of understanding. Providing immediate feedback within the live audio sessions is a distinctive feature of Centra. As previously noted, Centra sessions can be recorded and archived for future reference by instructors, students who participated in the session, and students who were absent from the session.

Blackboard enables frequent, timely feedback through a variety of functions: e-mail, discussion board, digital drop box, and test scoring and reporting. The communication function (see Fig. 7.2) in Blackboard, supports e-mailing all or any subset of students through the general e-mail option or the e-mail option for each group within the Group Pages function (see Fig. 7.2). The discussion board function (see Fig. 7.2) allows instructors and peers to respond asynchronously to comments and queries by individuals or groups of students. When students submit reports through the Digital Drop Box under Tools (see Fig. 7.1), instructors can respond with critiques that are comparable to comments on written papers submitted in a conventional, face-to-face course. The testing function of Blackboard scores student responses and posts the test score for immediate viewing by the student. Feedback on correct and incorrect responses for each test item enables students to correct misunderstandings. Thus, several Blackboard and Centra functions support feedback that enables students to be active monitors and regulators of their learning.

Principle 5: Emphasizing Time on Task

College students of the Net Generation prefer technology-based communication and rapid learning with immediate outcomes (Howe & Strauss, 2000). Extrapolating from Vygotsky's (1962) social constructivism, individual factors to be considered for promoting time on task include a student's prior knowledge, problem-solving skills, and motivation for a specified task. Has the student acquired the knowledge and skills necessary for proceeding with the task? Has the student developed the requisite study or critical thinking skills? Does the student believe the task has personal or heuristic value? Apart from generalizations derived from specifying prerequisite courses, instructors typically do not have individualized documentation for such questions. Cohort factors influencing time on task include age, experience, and disposition toward learning.

In spite of the aforementioned ambiguities, a variety of strategies and techniques are available for facilitating time on task in online courses. Traditionally, the concept of *time on task* is narrowly defined to mean the amount of time elapsed between the initial task involvement and the termination of involvement (Santrock, 2001). By expanding the concept to include timing and timeliness of student activity, instructors can implement the full range of online resources. With the focus on personal readiness for learning a particular task, timing addresses when students begin a task and the amount of time scheduled for that task. On this point, the asynchronous feature of online learning offers many advantages over conventional, face-to-face settings. *Timeliness* addresses the relation of the task to cognate concepts within the unit or course. The expanded set of functions for communicating with students in online courses equips instructors for assisting students to stay on schedule.

Blackboard offers a variety of asynchronous functions for facilitating both duration and quality of time on task in online courses. Prior to beginning a course, instructors might insure that students have adequate experience and expertise with the required technology through precourse surveys and practice modules linked to Course Documents (see Fig. 7.1). Communication through e-mail can convey the schedule of due dates for the entire course, posted in or linked to Course Information. Through consultation with peers or a select sample of students, instructors can verify the clarity and coherence of instructions and instructional goals. During the course, use the Announcements function when information is timely but not time sensitive such as recommendations for time management of a specified task. For time-sensitive information such as impending deadlines or commentaries on learning activities in progress, use the e-mail option under the Communication function to inform all students, select groups, or individuals. Facilitate improved quality of time on task by using the Assignment option in the Assignment function. The Assignment option requires students to enter a response to the assignment in order to exit the activity. Requiring an analysis and critique insures a thoughtful reading and comprehension of the assignment. The asynchronous Blackboard functions for promoting time on task are complemented by the synchronous nature of required Centra-based, discussion sessions.

Principle 6: Communicating High Expectations

The aggregate effect of implementing the prior five principles for good practice in undergraduate education tends to convey high expectations for performance in online courses. A consistent, high level of instructor involvement with online students exemplifies the interaction that is central to Vygotsky's (1962) social constructivism and meets the expectations of the socially oriented college students of the Net Generation. Requiring cooperative learning conveys the importance of personal and interpersonal responsibility for learning outcomes. Incorporating frequent opportunities for active learning is consistent with the desire for experiential learning among Net Generation college students. Providing frequent, task-specific feedback about students' understandings and misunderstandings tends to motivate them toward improved performance. Development tends to progress from novice to expert levels of knowledge and skills when students routinely practice quality time on task.

In addition to the implicit messages emanating from implementation of the seven principles for good practice in undergraduate education, instructors can convey explicit expectations about individual responsibility and performance. Prior to the start of the course, post a complete syllabus that clearly conveys equivalency between the online course and the same course offered in a conventional, face-to-face setting. (See a sample syllabus on the CD accompanying this volume.) Additionally, through introductory communiqués, instructors can address common misconceptions about online education. Online courses are not easier than conventional, face-to-face courses. Online courses are not student paced, lacking scheduled submissions and specified deadlines. Online courses are not correspondence courses, lacking interaction with instructors. Online learning is not student regulated, lacking feedback and mentoring by instructors.

Principle 7: Respecting Diverse Talents and Ways of Learning

The universality of access to online education dictates that designers of online courses be sensitive to diversity, in the broadest sense, in learning styles; in behaviors, attitudes, and expectations of students; and in degrees of visual, auditory, motor, and cognitive impairments. The Web-based technologies and pedagogical practices constituting the JMU model support a wide range of modes interaction and achievement. Text-based functions are integrated with complementary visual and auditory modes of interaction. Asynchronous and synchronous modes of interaction are maximized for both students and instructors. Student achievement is facilitated by an array of individual and small group experiential learning opportunities. Overall, the JMU model exhibits high levels of accessibility, flexibility, convenience, and adaptability in support of optimal student achievement.

SUMMARY

Implementing the JMU model for designing of an online introductory statistics course enables equivalency of structure and functions typical of courses in conventional, face-to-face settings. The model is applicable for a variety of online courses: introductory and advanced, concept-oriented and skill-oriented, lecture and lab-based courses. Vygotsky's (1962) social constructivism provides the theoretical framework for a pedagogy that features modeling, experiential learning, and interaction among students and instructors. Implementation of the theoretical framework involved integrating five Web-based technologies with Chickering and Gamson's (1999) seven principles for good practice in undergraduate education.

We believe it is appropriate to note that the JMU model is a dynamic concept, not a specific set of Web-based technologies or an immutable pedagogy. Only dynamic concepts can maintain continuing relevance with emerging learning environments supported by rapidly changing technologies (Jones, 2002). We encourage dynamic replications of the model in the context of one's institutional resources and support systems. The criterion for success of such efforts will not be the adoption of a specific set of technologies or the implementation of a particular pedagogy but, rather, the facilitation of optimal development of and achievement by our students, our *raison d'être*.

AUTHORS' NOTE

Charles M. Harris is a Professor in the Department of Psychology. James G. Mazoué is Assistant Professor of Educational Technologies and Dis-

tance Learning Coordinator in the Center for Instructional Technology. Hasan Hamdan is Assistant Professor and Arlene R. Casiple is Instructor in the Mathematics and Statistics Department.
Send correspondence to Charles M. Harris, Department of Psychology, MSC 7401, James Madison University, Harrisonburg, VA 22807; email: harriscm@jmu.edu

REFERENCES

Aronson, E., & Patnoe, S. (1997). *The jigsaw classroom: Building cooperation in the classroom.* New York: Longman.

Ayersman, D. J. (1995). Introduction to hypermedia as a knowledge representation system. *Computers in Human Behavior,* 529–531.

Benedict, J. O., & Anderton, J. B. (2004). Applying the just-in-time teaching approach to teaching statistics. *Teaching of Psychology, 31,* 197–199.

Bruning, R., Schraw, G., & Ronning, R. (1999). *Cognitive psychology and instruction* (3rd ed.). Upper Saddle River, NJ: Prentice-Hall.

Buxton, L. (1991). *Math panic.* Portsmouth, NH: Heinemann.

Chickering, A. W., & Gamson, Z. F. (1987). Seven principles for good practice in undergraduate education. *AAHE Bulletin, 39*(7), 3–7.

Chickering, A. W., & Gamson, Z. F. (1999). Development and adaptations of the seven principles of good practice in undergraduate education. *New Directions For Teaching and Learning, No. 80.* San Francisco: Jossey-Bass.

Duffy, T. M., & Jonassen, D. H. (1992). Constructivism: New implications for instructional technology. In T. M. Duffy & D. H. Jonnassen (Eds.), *Constructivism and the technology of instruction* (pp. 1–16). Hillsdale, NJ: Lawrence Erlbaum Associates.

Elbaum, B., McIntyre, C., & Smith, A. (2002). *Essential elements: Prepare, design, and teach your online course.* Madison, WI: Atwood.

Hackworth, R. D. (1985). *Math anxiety reduction.* Clearwater, FL: H & H.

Harris, C. M. (2005). *Conceptualizing an online social sciences course.* Innovative Teaching Strategies for Faculty Using Blackboard Conference. Virginia Commonwealth University, Richmond, VA.

Holmberg, B. (1989). *Theory and practice in distance education.* London: Croom Helm.

Howe, N., & Strauss, W. (2000). *Millennials rising.* New York: Vintage Books.

Jonassen, D. H. (1999). Designing constructivist learning environments. In C. M. Reigeluth (Ed.), *Instructional-design theories and models: A new paradigm of Instructional theory* (Vol. II, pp. 215–239). Mahwah, NJ: Lawrence Erlbaum Associates.

Jones, S. (2002). The Internet goes to college: *How students are living in the future with today's technology.* Washington, DC: Pew Internet & American Life Project.

Kuh, G. D. (1994). Student learning outside the classroom: Transcending artificial boundaries. *ASHE–ERIC Higher Education Report No. 8.* Washington, DC: The George Washington University.

Santrock, J. W. (2001). *Educational psychology.* Boston: McGraw-Hill.

Shank, P., & Sitze, A. (2004). *Making sense of online learning: A guide for beginners and the truly skeptical.* San Francisco: Pfeiffer.

Slavin, R. E. (1995). *Cooperative learning: Theory, research, and practice* (2nd ed.). Boston: Allyn & Bacon.

Velsmid, D. A. (1997). The electronic classroom. *Link-up, 14*(1), 32–33.
Vygotsky, L. S. (1962). *Thought and language*. Cambridge, MA: MIT Press.
Zaslavsky, C. (1994). *Fear of math: How to get over it and get on with your life*. New Brunswick, NJ: Rutgers University Press.

Chapter 8

Beyond Basics: Enhancing Undergraduate Statistics Instruction

Karen Brakke
Spelman College
Janie H. Wilson
Georgia Southern University
Dolores V. Bradley
Spelman College

Psychology departments across the nation recognize the importance of frequent review and update of curricular offerings. Teachers of content-area courses must change their lectures and syllabi to keep up with new research findings and new or revised theories that psychologists consider important to understanding human thought and behavior. Many psychologists do not recognize that the same process of review and update must be in place for statistics and methods courses, as well. The techniques used for data analysis, for example, have evolved through the development of decision-making processes that rely on logical reasoning as much as on the mathematical manipulation of numbers that undergraduates associate with statistics. Like any content area offerings, these courses must reflect current and ever-changing knowledge and argument of professionals in the field.

Recent decades have witnessed changes in the status quo of statistical analysis (see Kirk, chap. 3., this volume). The computational power of computer technology has allowed development of sophisticated programming and software packages that can perform highly complex analyses. Some of these analyses support the general linear model (GLM) and null hypothesis statistical testing (NHST) upon which psychology researchers rely heavily. However, the use of computers now facilitates other types of analyses—for example, nonlinear modeling and Bayesian analyses. In addition, investigators can conduct complements to NHST, such as power analysis and effect size measurement, much more quickly and easily than they could just a few years ago.

The upshot of these developments has been that students today must learn more about statistical analysis than students of earlier generations in order to be competent producers and consumers of analyses. Competitive graduate programs, and some employers, expect entering students or employees to be ready to tackle the complexities of new analyses and to be conversant with the issues facing the field (e.g., Cohen, 1994). At the same time, there is still heavy reliance on the NHST standards that psychologists have used for years, and students must have facility with these basic GLM techniques as well.

It is incumbent upon teachers of undergraduate statistics courses, then, to ensure that their graduates are prepared for these challenges. Courses must have the flexibility to incorporate new techniques and perspectives, while providing a solid foundation in the NHST techniques that long have been the core of statistics courses. Add these requirements to the usual challenges of student engagement and understanding that statistics instructors know all too well, and the demands placed on a one-semester introductory statistics course can become daunting.

In this chapter, we discuss two means to meet the goal of enhancing statistics education. The first illustrates how an instructor may streamline an existing one-semester course, based on a course taught by the second author (Janie Wilson; JW). The second approach we discuss provides an example of a second-semester course taught by the first and third authors (Karen Brakke; KB, and Dolores Bradley; DB). These are by no means the only viable approaches, and both bring with them their own challenges. However, they are strategies that we have used with some success; we share them here in hopes of providing ideas and stimulating conversation about going beyond the basics in undergraduate statistical education.

STREAMLINING A SINGLE-SEMESTER COURSE

A course in statistics is one of the backbone requirements of a healthy psychology major (Halonen & the Task Force on Undergraduate Major Competencies, 2002). A recent survey of colleges and universities indicated that 93% offered a statistics course, and 86% devoted the entire course to statistics; however, only 6% required undergraduates to complete a second statistics course (Freidrich, Buday, & Kerr, 2000). Of statistics courses required for the psychology major, 28% were offered outside of the Psychology Department.

For the past several years, Georgia Southern University has required psychology majors to take a statistics course in the Mathematics Department prior to enrolling in psychological statistics. This prerequisite might be considered beneficial in that students are more prepared for a subsequent course, but the reality of the situation is that students do not seem to be more prepared, often arguing that they did not learn *z* testing, *t* tests, and correlation in the math course; even descriptive statistics appear foreign to them. One potential explanation for this pervasive problem is that math statistics are not generally applied. In contrast, instructors teach statistics for psychology in the context of examples, and numbers begin to take on meaning. If application indeed enhances learning, statistics courses for psychology majors should be offered in the Psychology Department when possible.

Regardless of where a statistics course is taught, undergraduate statistics involve a great deal of information. A recent sample of statistics teachers reported teaching correlation; independent-samples *t* test; and one-way, between-groups ANOVA in a one-term course (Freidrich et al., 2000). Unfortunately, instructors often teach correlations only as descriptive and abandon correlations before introducing inferential statistics. Further, students rarely learn more complex ANOVAs and multiple regression in a one-term undergraduate statistics course.

As instructors try to include more topics in a one-term course, we must choose between two philosophies: cover the information more quickly or omit information that is not needed in most research designs. Because statistics tends to be difficult for many students, I (JW) choose to streamline the course as indicated in Table 8.1. By periodically removing information not likely to be needed by undergraduates, instructors can include higher level analyses in a one-term course. For example, when describing variables, I cover scales of measurement (nominal, ordinal, interval, and ratio), but I omit discrete versus continuous and qualitative versus quantitative distinctions. When presenting distributions, simple frequency figures remain important; tables are too simplistic, and students rarely encounter relative frequency or cumulative frequency in real life. When teaching correlation, I cover only Pearson's *r* as the most likely type of correlation students will encounter. Even with a streamlined course, I teach hand calculations as well as the popular software program, SPSS, for most analyses.

Within the term, students explore two additional topics without hand calculations. After covering related tests, students learn the one-way, repeated-measures ANOVA and multiple regression using SPSS. If time allows, I also explain the mixed-design ANOVA and show students how to type data into SPSS for one between-groups factor and one within-groups factor. For each of these three analyses, I bring sample output to class, and we work through relevant information, highlighting what they might need in the future.

In addition to deciding which topics to cover, we must also decide which approaches to include, such as effect size, confidence intervals, and other issues debated in the statistics community (e.g., Capraro, 2004; Wilkinson & the Task Force on Statistical Inference, 1999). In hypothesis testing, the .05 alpha level is arbitrary, and instructors recognize that rejecting the null hypothesis will happen with a large enough *N*. Effect sizes indicate meaningful relationships or effects rather than

TABLE 8.1

Proposed Schedule of Statistics Topics and Testing Based on a 16-week Term

Time Required	*Topics Covered*
3 weeks:	Overview of statistics and types of studies (correlational & experimental)
	Types of variables (nominal, ordinal, interval, & ratio)
	Simple frequency graphs
	General shapes of graphs (e.g., skew & kurtosis)
	Measures of central tendency (mode, median, & mean)
	Introduction to SPSS
	Measures of variability (range, deviation, standard deviation, & variance)
	Test #1
2 weeks:	Descriptive z scores
	Probability and the inferential z
	Inferential z test and hypothesis testing
	Test #2
2 weeks:	Single-sample t test
	Related- samples t test
	SPSS for related samples
	Independent-samples t test
	SPSS for independent-samples
	Test #3
3 weeks:	One-way, between-groups ANOVA
	Post-hoc tests (Tukey's HSD & Fisher's protected t tests)
	SPSS for one-way, between-groups ANOVA
	Graphing data from studies
	One-way, repeated-measures ANOVA
	SPSS for one-way, repeated-measures ANOVA
	Test #4
3 weeks:	Two-way, between-groups ANOVA
	Post-hoc tests
	SPSS for two-way, between-groups ANOVA
	Graphing main effects and interactions
	Test #5
2-3 weeks:	Correlation

Linear regression (multiple regression)

SPSS for correlation and regression

One-way χ^2 (goodness-of-fit test)

Two-way χ^2 (test of independence)

SPSS for two-way χ^2

Test #6

just significant ones. Huberty (1993) pointed out that textbooks in the 1930s called for measurements of magnitude, and Moroney (1951) argued that statistics should not take the place of common sense. Finally, in the 1970s, "effect size" became popular among psychologists (Huberty, 1993). Changes in reporting statistics have occurred in APA style as well. Hypothesis testing should be coupled with effect size. In fact, the *Publication Manual* argues that "failure to report effect sizes" is a "defect" in reporting (American Psychological Association, 2001, p. 5). Eventually, confidence intervals may be commonplace in the one-term undergraduate statistics course; however, wording in the *APA Manual* suggests that they are not as crucial as effect size. The manual indicates that "the use of confidence intervals" is "strongly recommended" (American Psychological Association, 2001, p. 22).

With so much information to cover, many instructors turn to a software package such as SPSS for computations. Indeed, professors should teach a statistics package to give students more power to analyze their data efficiently. The most widely used software is SPSS. According to Freidrich and colleagues (2000), 87% of students in their sample used a statistics package, and over half of the sample used SPSS.

Benefits of using a computer program are numerous. First, Gratz, Volpe, and Kind (1993) reported that students feel pride in their ability to use the computer effectively. Second, students and instructors have a similar core of knowledge, and we can help each other with analysis questions. Third, students who use SPSS are also more prepared for graduate school, if they choose to apply. Fourth, students can bring valuable skills to whatever job they might take, even if it is not related to psychology. For example, a former student contacted me from the registrar's office to say she analyzed archival data using SPSS. In fact, I regularly include information about SPSS in letters of recommendation to let graduate schools and potential employers know that students are proficient in data analysis.

It might be tempting to teach students data analysis using only a computer program; this approach would certainly allow time to cover more analyses in one term. Unfortunately, failure to cover hand calculations removes a potentially useful way to actually build theory. Over a decade of teaching statistics has also taught me that students feel a sense of ownership and involvement when they calculate a statistic on their own. Gratz and colleagues (1993) reported that students who learned statistics exclusively using SPSS did get more accurate numerical answers than students using hand calculations, but students taught with hand calculations tended ($p = .08$) to make higher scores on open-ended questions of statistics. This finding may mean that students who learned using hand-calculations tended to have a deeper understanding of statistics, whereas those taught using only SPSS were able to get accurate numbers but perhaps had less understanding of what those numbers meant. Thus, hand calculations help to solidify theory from which students can naturally progress to SPSS for efficient data analysis.

By the time students complete the single term of psychological statistics, I want them to be able to look at a design and data set and know how to analyze the data as well as make sense of what they find. To this end, beginning with the related-samples *t* test, students compile a workbook of data sets with at least one example of each type of inferential statistic. Each inferential statistic is tabbed to allow quick access, and students often create a table of contents to better organize their work. An example for the one-way, between-groups ANOVA follows (from Wilson, 2005, p. 167).

> *You want to figure out what kind of boy attracts high-school girls. High-school girls were randomly assigned to hear about a guy with trendy clothes, a nice car, or a confident attitude, then they rated him on a scale from 1–10, with 1 being "not interested" and 10 being "very interested." Here are the ratings given by your participants:*

Trendy Clothes	*Nice Car*	*Confidence*
8	6	10
6	7	9
9	8	10
6	6	10
8	4	8
6	5	9
7	5	10
6	6	9
6	8	8

In the student-created workbook, each example is followed by data layout in SPSS (printed from the spreadsheet) and data output from SPSS. Relevant information on the output is highlighted, and the statistical value is written in APA style. Next, students synthesize their results into a typed plain-English section. Finally, students create figures to illustrate their data when figures are relevant (i.e., all workbook examples except chi-square analyses).

The workbook serves as an excellent capstone experience for psychological statistics. After covering so many analyses in one course, students appreciate having this workbook to take with them to future courses and, often, on to graduate school. Several generous students have even contacted me from graduate school to let me know how prepared they were and that other students want copies of their workbooks.

Statistics teachers certainly have a great deal of information to teach, especially when limited to one statistics course in psychology. Ever-changing emphases in statistics, reflected in APA manuscript requirements, necessitates that instructors reassess courses on a regular basis. It is difficult to let go of comfortable lectures, but reexamining courses often means just that. Professors must streamline courses to make way for high-level statistics or to add new approaches to old analyses. Fitting so much information into one course, and doing so effectively, is a challenge. But then, we believe that conscientious teachers of statistics have always risen to the challenge of revealing the ease and usefulness of statistics.

CREATING A SECOND-LEVEL STATISTICS COURSE

Another option for supplementing basic statistics instruction is to develop a second-level course for undergraduates. This option typically involves more effort than streamlining an existing course. Additional space, time, and staffing must be available, and often one must gain approval for a course proposal from an institutional or departmental curriculum committee. However, such a course can offer many benefits. Along with the additional time that students have to learn and practice statistical analyses, a course that supplements standard instruction carries with it great flexibility, both in terms of material covered and instructional techniques. Teachers may tailor the course to the needs and interests of students at different institutions, preparing them in many practical ways for graduate school or the world of work. One of the great delights of teaching an advanced course in statistics is that you have the opportunity to teach students about the joy of discovery. They often view statistics as a necessary evil and fail to appreciate the intellectual appeal of it as tool that one uses to answer questions. Introductory courses rarely have the time to address such concepts adequately; many students leave an introductory course believing that performing a statistical test is all that one does to a set of data, after calculating the mean and standard deviation. In a second course in statistics, however, students learn that a statistical test is only a part of the analytic process. Therefore, an instructor's goal is not only to have the students become better at performing descriptive and inferential statistics, but also to become proficient in integrating statistics with interpretation and communication.

An undergraduate psychology instructor must design a second-level statistics course carefully, keeping in mind the population served by such a course. Potential enrollees have some familiarity with statistical analyses, but they remain undergraduates. As such, they are not necessarily confident or convinced that statistics is "for them," although if the course is an elective, motivated students will tend to self-select through their enrollment. Thus, we (KB & DB) have found that building students' confidence in their abilities to be one of the primary aims of the course that we offer and that this confidence emerges, not surprisingly, through providing a balance of supportive instruction and challenges to think and work through problems on one's own.

The course schedule for a second-level statistics course is not obvious; in fact, it can go in many directions. Textbooks aimed at this level of undergraduate student are relatively rare. In our experience, the available texts tend to either be introductory, aimed too "low and slow" for these students, or they are rather "high and dry," written for graduate students and not designed to promote engagement among undergraduates who want to be excited about statistics but are not yet quite convinced that they will be.

Fortunately, although it may mean extra work and planning for the instructor, the scarcity of appropriate prepackaged curricula also affords opportunity to customize the course with interesting assignments and topics that fit the students' competencies as well as their future plans. For example, an instructor who has several students who intend to pursue PhD degrees in psychology might design a very different course than one whose students lean toward public health or market research careers. With a little creative thinking and some perusal of the many available print and electronic resources available to aid statistics instruction (some of which are listed on the CD accompanying this volume), one can develop a highly engaging, instructive, and relevant course.

We offer here discussion of one example of a second-level statistics course that we have developed over a number of years. We have offered this Statistics II course at Spelman College annually as a departmental elective, taught by either KB or DB. Recently, however, its status has changed, and it is now one of two options that students may take as an advanced measurement/analysis course (the other is Psychometrics). With the increasing demand accompanying this status change, the department offers the course every semester so that class size remains relatively small (i.e., under 25). Students take the course during the junior or senior year, following a required introductory statistics class that they typically take during the sophomore year. Students usually have also completed a required experimental design course that includes a review of basic statistical concepts. The outline of the most recent offering of the course appears in Table 8.2, although we encourage instructors to develop schedules that fit their students' needs and interests.

The main goals for the course are threefold. First, we try to strengthen the students' skills with elementary statistics, as well as

TABLE 8.2
Outline of Second-Level Statistics Course

Week	*Topics Covered*
1	Review measurement and graphics
2	Review descriptive statistics
3	Review hypothesis testing
4	t-tests (Project 1 assignment)
5	Power and effect size
6	ANOVA: Single factor; post-hoc tests
7	ANOVA: Factorial designs
8	ANOVA: Advanced technique (Project 2 assignment)
9	Correlation
10	Simple and multiple regression (Project 3 assignment)
11	Multiple regression
12	Chi-square (Project 4 assignment)
13	Bayesian approach to statistics
14	Wrap-up; supplementary topics (Project 5 assignment spans the semester)

their flexibility in applying and understanding basic statistical theory and practice. Second, we introduce additional analyses, focusing on multivariate GLM techniques but also providing brief coverage of simple Bayesian concepts as an alternative to NHST that public health professionals and others often use (see Gigerenzer & Hoffrage, 1995, and Kurzenhauser & Hoffrage, 2002, for a student-friendly means of teaching this material). With the new material we take a survey approach, providing enough coverage so that students can understand a journal article using one of the techniques or conduct simple analyses of their own. Finally, and perhaps most importantly, we try to instill in our students the confidence and adaptability that will serve them in later endeavors when they must conduct analyses in different contexts and under different constraints. By being willing to try and knowing how to make use of appropriate resources when facing a research problem, we believe that our students will be prepared to enter graduate school or professional life.

To that end, the course has an applied focus, although we often make reference to the underlying theory and reasoning of statistics. The instructional components that we include across nearly all topics are SPSS practice, interpretation of analyses, design and completion of dataset analyses, and collateral exercises involving such things as power analyses and codebook design. Students complete assignments after nearly every class, and we make every attempt to provide prompt feedback on them. We have discovered several assignments that work for us, and we present some of them here.

Course Pretest

There is nothing better than getting a lot of "bang for the buck" out of an exercise, and the course pretest has provided a lot of "bang." One of us (DB) developed this assessment for the advanced statistics course in order to get a sense of students' experience and knowledge upon entering the class. The pretest consists of 35 terms and concepts that students must match to their appropriate definitions (a copy of the pretest appears on the CD accompanying this volume). It includes items such as *inferential statistics, random sample, median, skewed distribution, repeated measures, and Type II error,* along with a variety of statistical tests. We use the results of the pretest to determine what areas need review at the start of the term. Where we get "more bang" is by having students use the results of the pretest to review the concepts that they learned in the first course in statistics. The students receive a matrix of the results of the pretest, which lists the question and the responses for each, anonymous, respondent. Students enter the data into SPSS and perform a series of descriptive statistics, including mean, standard deviations, frequency distributions, class intervals, percentiles, and so on. They are then to submit an APA-style typed summary of the data, with all tables and graphs having appropriate labels, captions, and format. As the first assignment of the course, it sets the tone for the high expectations; they have had the introductory course, the advanced course is for the serious preprofessional. The students enjoy analyzing a dataset to which they contributed, which is usually a new experience for them. Because of this personal involvement, they tend to approach the task with greater interest, which eliminates any apprehension they may have concerning their ability to recall what they learned in the introductory course.

The pretest is now a standard for both of us (KB and DB). We have extended its use by re-administering the test in the middle and at the end of the semester, with the expectation that the posttest results will be quite different from the pretest. We often supplement data from the test itself with other information such as how long it has been since each student completed the introductory statistics course, how long it takes to complete the assessment, estimated scores, and so on. Students receive a matrix of the data and perform descriptive and inferential analyses, which they report in detailed APA-style, Results and Discussion sections. They must select and perform the appropriate analyses and include an appropriate graph to go along with each type of analysis. They are then to answer the question "What can you conclude from your analyses?" Again, it is surprising how eager students are to work with data that they generated. Working with the pretest and posttest dataset is also an enjoyable and useful activity for the instructor because it is a measure of how far students have come since the start of the course.

Online Resources

Other assignments that "work" have often been adaptations from online sources, such as Web Interface for Statistical Education (WISE: http://wise.cgu.edu/index.html) and the Data and Story Library (DASL: http://lib.stat.cmu.edu/DASL/). These and other sites provide datasets, interactive graphs and calculators, and other exercises that engage the students while illustrating statistical concepts. For example, as described in the next section, the advanced statistics students serve as consultants for students who conduct research projects in other courses. The students conduct a "test run" for this project by completing a modified version of the WISE tutorial entitled *Choosing the Correct Statistical Test* (http://wise.cgu.edu/choosemod/opening.htm). The online test presents a research design and asks the respondent to select which of four statistical tests is appropriate. In the paper form that we give to the students, they must also explain why they chose a particular test. Correct answers bolster their confidence in serving as a real-life consultant. Incorrect answers reveal areas that need further study by the student.

Another online resource that we find particularly useful is the DASL Web site, which contains real-world datasets for a variety of disciplines and occupations. The section for each study includes an abstract, the original journal or text reference, variable names, and a dataset. The instructor may easily develop assignments to analyze the data, using descriptive and inferential analyses, and submit a detailed Results section and a well-constructed graph, which will answer the question posed by the study. Thus, for example, using the dataset that concerns refusals in mortgage lending rates, from the Association of Community Organizations for Reform Now (ACORN; 1992, available from http://lib.stat.cmu.edu/DASL/), students must determine whether ACORN's conclusions regarding racial disparities in mortgage lending are accurate. By using such real-world datasets, students can see that statistics are applied in many occupations. Moreover, not surprisingly, students often want to know the answers to the questions, so they tend to put in the time and effort to submit a high-quality assignment.

Project Applications

Projects are an essential component of the course. In fact, one of us (KB) has eliminated examinations entirely in favor of major project evaluations. Under this design, students must complete five major projects during the semester in addition to daily homework exercises. Three of these projects involve dataset analyses that require the students, either individually or in pairs, to take on the role of statistical consultants to an organization or research team that needs assistance. Datasets are readily available on the Web; we regularly use ones from the Data and Story Library and PsychExperiments (http://psychexps.olemiss.edu; see also http://opl.apa.org). Along with their reports and analyses, students submit a reflective journal relating how they went about addressing the issues involved and what they learned from the project. Students also complete a project in which they review and critique a journal article, focusing on analyses, and then prepare a mock exercise based on those found in one of their workbooks (Holcomb, 2004). All of these projects tap into analytic skills that are needed in real-world situations and provide assessment opportunities based on appropriateness and thoroughness of the work performed, reflecting student learning in the course.

The final project incorporates a service component that we have developed for the course. The initial assignment was for each student in the class to serve as a statistical consultant on a research project in the Social Psychology course. Most students in social psychology have had Introductory Statistics 1 to 2 two years previously, but may have forgotten many of the details of completing statistical analyses or are unsure about whether they are proceeding appropriately. They benefit from guidance in helping them think through the issues and decisions that they face in completing their research projects. By pairing these students with consultants from the Statistics II course, a "win-win-win" situation emerged. Social Psychology students ended up with better projects than they would have completed on their own, advanced statistics students gained confidence during an opportunity to apply their fledgling expertise, and the instructors were spared having to answer some of the many requests for help from students that they encounter whenever research projects are due.

In light of the success of the consultant project, and because the size of the advanced statistics course has increased while demand for consultants has remained relatively constant, we have added additional options over the years. As an alternative to the social psychology consultant project, students may serve as lab assistants for the Introductory Statistics course or as tutors for individuals who are struggling in the introductory course. They may also help one of the college offices, such as Assessment or Institutional Research, with analyses of specific data sets. If any of these projects fall through, a student may complete a back-up assignment of serving as a judge at a local student research day instead. This option provides a good alternative for students who have trouble meeting with their consultees or tutees and who do not believe that they have a gradable project otherwise.

All students must keep a journal of their activities, including meeting dates (and attempts), content of discussion, questions that consultees had, and so on. They must also submit a reflective overview of the project at the end of the semester, in which they give an overview of the material in their journal, as well as provide their reactions to participation in the project. Students who participate in consultant projects may also fill out a Likert-scale evaluation form assessing their experience. Generally, the responses of the consultants and their consultees show good agreement on items regarding time spent and usefulness of the exercise, and their perceptions of this project, as well as others, are usually highly positive. Although consultants do not earn grades based on the performance of their advisees, we do assess them on the quality of their advice and the depth of thought put into their work.

CONCLUSIONS

Using course examples from Georgia Southern University and Spelman College, we have (a) outlined a way to streamline a one-term course to include moderate-level analyses and current updates in the field, and (b) reviewed one approach to a second-level statistics course that allows students to apply their knowledge in different contexts. The assignments we have discussed are consistent with the promotion and assessment of student learning outcomes, in that they result in document-based bodies of work that exemplify facets of statistical analysis. Whether taking the form of a comprehensive workbook, pre- and postcourse examinations, or a series of independent projects, the materials give evidence that the students are able to produce appropriate statistical analyses and understand statistical concepts. Instructors or departments can conduct further assessment by requesting external evaluation of service projects or by monitoring performance in subsequent courses that require statistical competence.

Although we trust that these approaches provide students with a firm foundation in statistics, we are all continually developing our courses in response to feedback from students, faculty colleagues, and others in the field. In addition, we encourage our colleagues who teach other courses in the department to incorporate statistical analyses whenever possible, thus keeping the concepts and skills fresh for students. We know that learning, as well as teaching, statistics is a challenge, but one that can be very rewarding when students become competent, confident participants in data analysis.

REFERENCES

American Psychological Association. (2001). *Publication manual of the American Psychological Association* (5th ed.). Washington, DC: Author.

Association of Community Organizations for Reform Now (1992). *Banking on Discrimination: Executive Summary, October, 1991, in Joint Hearings Before the Committee on Banking, Finance, and Urban Affairs, House of Representatives, 102nd Congress, 2nd Session, Serial Number 102-120* (pp. 236–246) [Data file]. Available from the Data and Story Library Web site http://lib.stat.cmu.edu/DASL/

Capraro, R. M. (2004). Statistical significance, effect size reporting, and confidence intervals: Best reporting practices. *Journal for Research in Mathematics Education, 35*, 57–62.

Cohen, J. (1994) The earth is round ($p < .05$). *American Psychologist, 49*, 997–1003.

Fredrich, J., Buday, E., & Kerr, D. (2000). Statistical training in psychology: A national survey and commentary on undergraduate programs. *Teaching of Psychology, 27,* 248–257.

Gigerenzer, G., & Hoffrage, U. (1995). How to improve Bayesian reasoning without instruction: Frequency formats. *Psychological Review, 102,* 684–704.

Gratz, Z. S., Volpe, G. D., & Kind, B. M. (1993). *Attitudes and achievement in introductory statistics classes: Traditional versus computer-supported instruction* (Rep. No. JH940166). Ellenville, NY: Teaching of Psychology: Ideas and innovations: Proceedings of the 7th Annual Conference on undergraduate Teaching of Psychology, March 24-26. (ERIC Document Reproduction Service No. ED365405)

Halonen, J., & the Task Force on Undergraduate Major Competencies (2002). *Undergraduate psychology major learning goals and outcomes: A report.* Retrieved from http://www.apa.org/ed/pcue/taskforcereport2.pdf

Holcomb, Z. (2004). *Interpreting basic statistics* (4th ed.). Glendale, CA: Pyrczak.

Huberty, C. J. (1993). Historical origins of statistical testing practices: The treatment of Fisher versus Neyman-Pearson views in textbooks. *Journal of Experimental Education, 61,* 317–333.

Kurzenhauser, S., & Hoffrage, U. (2002). Teaching Bayesian reasoning: An evaluation of a classroom tutorial for medical students. *Medical Teacher, 24,* 516–521.

Moroney, M. J. (1951). *Facts from figures.* New York: Penguin.

Wilkinson, L., & the Task Force on Statistical Inference. (1999). Statistical methods in psychology journals: Guidelines and explanations. *American Psychologist, 54,* 594–604.

Wilson, J. H. (2005). *Essential statistics.* Upper Saddle River, NJ: Pearson Prentice-Hall.

Part IV

Emerging Approaches to Teaching Research Methods

Chapter 9

Hands-On Labs in Content Area Methods Courses

Ruth L. Ault, Margaret P. Munger, Scott Tonidandel,
Cole Barton, and Kristi S. Multhaup
Davidson College

INTRODUCTION
RUTH L. AULT

The many subdisciplines of psychology favor different research methodologies, each with corresponding strengths and weaknesses. At Davidson College, we want to expose students to this range of techniques while we provide many hands-on experiences and more depth about each method than a single course can typically accomplish. We therefore offer multiple courses that emphasize different research methodologies but, at the same time, reinforce aspects that cross subdisciplinary boundaries (e.g., operationally defining independent and dependent variables, telling a clear story about the research one has conducted, writing in appropriate professional style). This chapter describes 4 of the 11 research methods courses that we offer, all with hands-on laboratories in particular content areas. To put that information in context, this section (a) gives a brief introduction to the college, (b) presents an overview of the requirements for our major, (c) emphasizes the role that research methods courses play, and (d) highlights what the courses have in common so that the other chapter sections can emphasize the unique aspects of these four courses.

Davidson College is a highly selective, residential, liberal arts undergraduate college of about 1,700 students, of whom 30 to 50 graduate each year with Psychology majors. We have 8.5 faculty with specialties in social, clinical, developmental, industrial/organizational, cognitive aging, perception, behavioral neuroscience, and psychopharmacology. The semester is 15 weeks long, and faculty members teach five courses a year, with laboratory courses typically counting as two classes.

The department mission statement includes a focus on research methodology to (a) contribute to new knowledge in the discipline; (b) make students more informed consumers of research findings; (c) prepare them for graduate study, not only in psychology, but in any discipline that requires scientific critical thinking; and (d) enhance their general educational experience by giving them a scholarly partnership with faculty members. (See www.davidson.edu/academic/psychology/major.htm or the CD accompanying this volume for the full statement.)

We thus designed a curriculum that requires courses of particular types as well as courses that make students sample different areas of the discipline. Specifically, the major requires 10 courses: General Psychology, three electives that students usually take from the broad survey courses available, Research Design and Analysis (our statistics course), three additional research methods courses, a seminar, and a capstone experience. The area requirements dictate that students have one course in each of four areas: animal-physiological, cognitive, developmental, and social–clinical–industrial/organizational. The complexity of four areas and three research methods courses per student means we need to offer a lot of research methods courses. In fact, we have 11 such courses, although not every one is offered every year.

Table 9.1 lists the titles of the research methods courses that fit under each of the four areas. One additional restriction is that of the three research methods courses, one must be from the animal–physiological or cognitive areas; one from developmental or social–clinical–industrial/organizational; and the third can be from any area. Students can use these research methods courses to help satisfy the overall area sampling, and many students do this–they take their three methods courses in three different areas and either a survey course or seminar in the fourth area. Enrollments in the courses are limited to between 14 and 18 students.

TABLE 9.1

Research Methods Course Titles that Meet the Four Area Requirements

No Statistics Prerequisite		*Statistics Prerequisite*	
Animal-Physiological	*Cognitive*	*Developmental*	*Social/Clinical/Industrial-Organizational*
Behavioral Pharmacology	Perception and Attention	Child Development	Social
Behavioral Neuroscience	Memory	Adult Development[a]	Clinical
Learning			Industrial-Organizational
Animal Behavior (taught in Biology Dept.)			

[a]No statistics prerequisite.

What do these courses have in common? The most obvious characteristic is that they all give students hands-on experience doing research. Students read primary sources such as journal articles, they get involved in designing a piece of research, they act as experimenters to collect data, they analyze the data, and they disseminate the results. In most of the courses, teams of students work together on at least one project; in many of the courses students also generate their own research proposal. Thus they are very busy over the 15 weeks of the semester! We hope for a mix of juniors and seniors, and we certainly have a mix of students for whom our methods course is their first, second, or third such experience.

Each course focuses on those methods that are common to that particular field of psychology, be it single-subject or group designs; survey, interview, observation, or experimental techniques; in the lab or in the field; with humans or nonhumans. About half of the courses also try to teach a broad sampling of the area's content; the other half focuses the content on one or a few topics. Some of the faculty assign a traditional journal article-style paper to disseminate results, some assign a poster presentation, and some have both papers and posters.

Each course also has unique elements that we highlight in the subsequent sections. In Margaret Munger's perception and attention class, students experience being both researcher and participant. Scott Tonidandel's description of the industrial/organizational course illustrates applied problems and settings as the laboratory. In the clinical course, Cole Barton emphasizes a scientist–practitioner model. Finally, Kristi Multhaup describes how she combines the adult development research methods course with a lower level psychology of aging course to the benefit of students in both classes.

PERCEPTION AND ATTENTION LAB: BEING PARTICIPANT AND RESEARCHER
MARGARET P. MUNGER

I teach Davidson's methods class in perception and attention to 16 students; the only prerequisite is General Psychology. In thinking about how to bring the area alive for the students, I realized that the students have trouble connecting method to data and, subsequently, data to research literature. In particular, I wanted students to learn to focus on the experimental task, link the data back to the experimental conditions, and use the literature to figure out the appropriate data analysis. I decided that one way to get students to begin thinking about the task, and how the very task itself is important, would be to have the students actually participate in experiments. We then analyze the class data, and they know that they are one of the participants and that those numbers correspond to something they did. I take advantage of the fact that many perception and attention studies do not require large samples, which lets me use class data, but I think this teaching method would still be meaningful if students analyzed a larger data set.

Each week we have a new topic; I have included on the companion CD the stimuli and data sets from two of my labs: mental rotation and facial attractiveness. Students have access to my lab, and each experiment includes computerized instructions allowing self-scheduled participation taking from 5 to 30 min. For most of the labs, the computer automatically collects the data, although sometimes the students have to e-mail it to me. I spend about 1 hr organizing the text files into an Excel® workbook, which I then distribute to the students as their starting point. Excel® is available in all our public labs at Davidson, so I use it for basic analysis and graphing. Each week we have a day in a computer classroom where I show the students around the Excel® workbook, which includes separate sheets for each participant and room for summary statistics and graphs. You can see how I format the Excel® workbook in the CD examples ("MR example.xls" and "face example.xls"). Class begins with a discussion about the experience of doing the experiment, and I reveal the experimental conditions. In the mental rotation lab, the object appeared rotated about one of four axes, whereas the facial attractiveness lab presented faces of varying thinness (see the corresponding "readme" files on the companion CD for full details). As students work with the data, I have them match the various data cells with what they did, we talk about what they thought they were doing, and we see if the numbers support their intuition.

Students learn how to code the data for each sheet and then "drill through" to make a summary table on a new sheet named *Summary*. Then, they are ready to calculate some statistics. Because statistics is not a prerequisite, I teach them what a confidence interval is and then do a quick explanation of a *t* test and how to use it. I spend a lot of time working on graphing—how to make an effective graph, not just a colorful one—and trying to get students to think about how a different graph of the same data can highlight a particular point. The CD with this volume includes the PowerPoint® I use to introduce the mental rotation data analysis.

The final product each week is a two- to three-page lab report that includes a brief introduction, method, results, and discussion section. Students must include appropriate graphs and eventually learn that they should not include all the graphs they make. For these weekly labs, students do not do additional library research, but just use three to five assigned journal articles for each lab; at least one of these articles has methods closely related to the particular task they have done. For the mental rotation lab, I have students read Shepard and Metzler (1971) and Murray (1997). For the facial attractiveness lab, I assign Rhodes et al. (2001) and Rhodes, Jeffery, Watson, Clifford, and Nakayama (2003). I provide more background in the lectures, generally from articles not assigned for reading, and then we discuss the assigned articles.

I have structured the sequencing of data analysis and articles two ways, and I think there are merits to both approaches. My first approach was to do all the lectures and readings before students saw the data, with the idea that an understanding of the background would help with data analysis. However, this structure meant that lecture and discussion were running half a week ahead of the reports they were writing, and students found it difficult to focus on the articles for class. With this approach, I would get a lot of good questions in office hours—about the previous week's material. That situation inspired my other approach of performing the data analysis on the first day of a new topic. With this approach, students have already participated in the lab and read the journal article that matches the design most closely. Then we spend the next two days in lecture and discussion reading additional articles and discussing related issues. I tend to end the section with a discussion of all of the related articles on the last class day before the report is due. This second scheme has led to much better discussion, although students occasionally started out pretty confused during the data analysis with a new topic.

There are some experiments where the data fit with students' expectations as participants, which can help students make links between task and data. Mental rotation is a nice example: When asked to make judgments about rotated objects, most individuals think they are mentally rotating a image, and their reaction time data increases with larger orientation differences, in keeping with their intuition (Shepard & Metzler, 1971). However, students also need to take the data into account as they interpret them, and so I do set up some of the labs as surprises. For example, when you consider simple stimuli, an abrupt onset will capture not only your attention, but your eyes, too (Theeuwes, Kramer, Hahn, & Irwin, 1998). Students who have read Theeuwes et al. will expect easy detection of an abrupt onset of an object in a scene. It is, however, very hard to find the changing item in a realistic scene, a phenomenon called *change-blindness* (Rensink, O'Regan, & Clark, 1997), so their background knowledge can help them see how surprising change-blindness results are.

There is a danger with using class data for analysis because sometimes we fail to replicate the phenomenon or we even find contradictory results; but when that happens, I make the students try and figure out why. When I push them to develop an idea beyond "we need more people," they make some very interesting points, which lets me see that they are finally thinking about the stimuli and the task in a truly meaningful way.

INDUSTRIAL/ORGANIZATIONAL: TEACHING METHODS IN APPLIED SETTINGS SCOTT TONIDANDEL

Often, you may want to teach research methods outside of a traditional laboratory environment, maybe because you do not have the necessary lab resources or perhaps the research questions of interest lend themselves more readily to applied methodologies. In either case, you could assemble a number of activities for your students to give them the necessary research methods experiences. In the following sections, I describe four research methods projects that I have used successfully in my Industrial/Organizational research methods class. For each activity, I discuss potential data sources and the methodological principles that each project illustrates.

Survey Design Project

A *survey design project* is an extremely versatile activity that can expose students to virtually all aspects of research. Students work in small teams on separate projects. Usually, I look to local businesses as data sources. The focus of the surveys may be issues such as customer service perceptions or patrons' preferences for different bank services. For one of the more original projects, a team examined music preferences (live music vs. disk jockey) at a local establishment. Local businesses are frequently willing to participate, especially if you offer them something in return for their participation. As a result, I require all of my research teams to give a separate business quality report to the organization upon completion of the project. A second source of data for this project can be organizations on campus. The physical plant staff, library employees, or information technology group may all be willing participants in surveys designed to ask pertinent human resource questions. For example, students interested in organizational commitment used hall counselors and residence life staff as their participants; others explored compensation issues using work–study students.

The survey design project is particularly useful because it involves students in the entire research process from start to finish. I require students to formulate their own research question, generate items that comprise the survey, develop a sampling strategy, and collect their data. Once students gather data, they perform detailed item analyses and test their hypotheses. They then produce both a written and oral presentation of their findings. In addition to enhancing students' understanding of research, this project helps develop practical consulting skills because students have to make contact with the organizations themselves and sell their research idea to the organization in order to get participation. Students are engaged in this project because they have flexibility in choosing a topic that interests them.

Personnel Selection Exercise

One area of industrial/organizational psychology that is extremely useful for demonstrating a variety of statistical techniques is personnel selection. Unfortunately, convincing an organization to open its selection process to a group of outsiders, much less a research methods class, is nearly impossible. Fortunately, college and university admissions processes produce copious amounts of selection data that a methods class may use. Davidson's admissions office has been more than willing to share multiple years of anonymous admissions data. These data sets may contain many useful variables such as SAT scores, high school GPA, numerical evaluations of application essays, and letters of recommendation. In addition, these data may be easily linked to freshman year grade point average (GPA), GPA at graduation, or student retention. All that the project requires are a couple of predictors and at least one relevant criterion variable to create a meaningful selection exercise.

I use these data sets primarily to illustrate the principles of personnel selection and validation to my students. Additionally, the selection exercise is particularly useful for demonstrating some advanced statistical techniques such as multiple regression, incremental validity, handling missing data, restriction of range, and identifying bias in a selection system. Students tend to really enjoy this particular exercise because the data and research questions are so salient to their lives as students.

Mini Meta-Analysis

A third project you can integrate into a research methods class is a mini meta-analysis. The word "mini" is important to stress here because anyone who has ever conducted a meta-analysis knows what an unwieldy task it can become. A key to successful implementation of such an activity is to keep it limited in scope. Perhaps the greatest advantage of the meta-analysis project is the availability of potential data sources. You need only a set of articles on a certain topic that contains necessary statistical information for meta-analytic computations. I like to use meta-analysis to introduce students to the issues of validity generalization but the techniques can be applied in any substantive area.

First and foremost, a meta-analysis project exposes students to the importance of replication in research and to the procedure of meta-analysis as a technique for cumulating research findings. In addition, students gain a much better understanding of issues relating to statistical power, type I error, effect size, and sampling. The meta-analysis project can take on varying levels of sophistication depending on the aims of a course. For instance, students can develop their own coding scheme for articles, examine issues of interrater agreement and interrater reliability, apply corrections for unreliability in the original articles, and identify potential moderator variables.

Training Evaluation

A fourth applied methods project that lends itself readily to implementation on academic campuses is training evaluation. Numerous groups on campus frequently perform training and would be willing to open their training programs for evaluation by a methods class in order to have some evidence of efficacy. A school's human resources department may be periodically engaged in training employees. Information technology groups often conduct ongoing software or computer training classes for students, faculty, and staff. Perhaps the richest source of potential training data is educational interventions in other classes. Colleagues may have grants to implement a new technology in the classroom, or they may be incorporating a new pedagogical tool into a course. In either case, they frequently would love to have some data regarding the intervention's efficacy that they can take back to the granting agency to request additional funding. Before the semester begins, I try to identify potential training evaluation opportunities on campus, and then I assign research teams to each one of the projects.

The training evaluation project involves students in the entire research process from start to finish. One of their first steps is to develop an evaluation design. Developing a design forces students to deal with threats to the validity of their design that are often less of a concern in well-controlled laboratory research. They have to decide what to measure, how to measure it, and how to create appropriate instruments for their constructs. Students are also responsible for collecting the data, analyzing the data, and presenting their findings in a written report and orally to the entire class. Again, students get experience with specific research techniques, but they are also exposed to the trials and tribulations of trying to conduct research in applied settings.

Summary of Teaching Applied Methods

Each of the four projects exposes students to a variety of methodological techniques in settings outside the conventional laboratory. These activities are flexible, allowing for various combinations of projects that can be done individually or in research teams. My approach is to use either the survey design project or the training evaluation project as the focal activity and supplement that experience with one or both of the remaining projects. Instructors can adapt the level of sophistication required for each project to meet specific course objectives and to ensure that a particular combination is feasible in a semester. In addition to providing students with experience conducting research, these activities afford additional benefits of exposing students to real-world dilemmas and fostering the skills needed to apply scientific principles in applied settings.

CLINICAL RESEARCH METHODS: EMPHASIS ON SCIENTIST–PRACTITIONER

COLE BARTON

Not many students become clinical psychologists because they are interested in clinical research. I orient students by asserting the fundamental importance of clinical research to the development and sustained professionalism of the field. To emphasize the relevance of research for even those who might not envision themselves as clinical professionals, I direct their attention to the American Psychological Association's (APA) Code of Ethics (American Psychological Association, 2002), where students quickly discover that they should use only assessments and interventions with demonstrated empirical utility. I point to contemporary changes in the health care milieu increasing clinicians' accountability for the demonstrated efficacy of their interventions.

Ecological Model and Clinical Methods

The course surveys the methods and paradigms in clinical psychology as cast within a conceptual scheme originally devised by Bronfenbrenner (1977). As shown in Table 9.2, each ecological dimension of a level of analysis has corresponding methods and questions associated with it.

I expect students to have exposure and contribute to the development of each of four course projects, and I assign students to a team with principal responsibility for writing up and reporting one of the projects. I assign teams based on students' declared interest in a given project, but I try to also balance the teams with my perceptions of their technical competencies with statistics and equipment, as well as my informal appraisals of their responsibility and leadership styles. The team that is principally responsible for a given project develops the work plan accommodating the other teams' rotations through their project, providing oversight and management. Teams then report to me on a scheduled basis, and we use project management sheets to evaluate how well the

TABLE 9.2
Ecological Model and Clinical Research Methods

Bronfenbrenner Ecology	*Representative Project and Methods*
Ontogenetic	Psychophysiology Analogue Study
Property of an individual	Arousal, expectancy, and alcohol consumption
Microsystem	Observational Study
Important dyadic relationship	Process coding physician-patient interaction
Exosystem	Self-Report Survey
Community influences on behavior	Attitudes toward alcohol and an institutional policy on its use
Macrosystem	Experimental Survey Method
Cultural influences on behavior	Perceptions of mental illness as a function of behavioral descriptors

project is moving to the finish line. Finally, at the end of the semester, teams deliver a PowerPoint® presentation to the class as well as an APA-format journal manuscript.

Clinical Studies Without Data Collection

It can be prohibitive to collect clinical data for a host of ethical or logistic reasons. Alternatively students can develop clinical research skills by analyzing textbook materials.

Controversies. Several excellent supplemental texts in abnormal or clinical psychology review pro and con positions on important controversies in the field (e.g., Halgin, 2005). Although these texts can motivate students to analyze the primary empirical literature, the authors aim their excellent overviews at the introductory level student.

Students taking my course have had two course prerequisites (abnormal psychology and statistics). Students look for more detailed and contemporary articles supporting the pro or con position in the supplemental textbooks, or they critically review some of the primary sources in the articles for more sophisticated analysis of the arguments. For example, one group of students conducted a meta-analysis of the effectiveness of an intervention model applied to substance abuse. To move students beyond just methodological critique, I make up or adapt data sets where students conduct statistical analyses modeled on the controversies and ask them to write a brief report summarizing how their analyses might provide a resolution.

Clinical Write-Up. Many excellent case studies serve as enrichment supplements for abnormal psychology courses (e.g., Gorenstein & Comer, 2002; Halgin & Whitbourne, 1998). Students read a case study with obvious labels and identifiers edited out. They must formulate a diagnosis and treatment plan within *Therascribe* (Jongsma, 2001), a clinical case management software that is essentially a relational database. Once students have some preliminary ideas from their reading of the case, the software prompts them to make a host of decisions about the case, walking them through the steps for incorporating data from assessments, making a diagnosis, developing a treatment plan, setting treatment goals, and evaluating an outcome. Because Therascribe collects data within a relational database, students can summarize the entire class's data and perform epidemiological or outcome studies on those pseudodata as well. In the course of these exercises I hope to establish parallels between clinical decision making and the design and interpretation of good research. The clustering of symptoms into a syndrome is akin to the structure of convergent validity, and clinical "rule out" decisions parallel discriminant validity. I point to examples such as the temporal consistency in waxing and waning mood for bipolar disorder, with parallels to test–retest forms of reliability. We discuss issues such as interobserver reliability in clinical work and evaluating clinician agreement for questions of gender or ethnic bias in diagnosis and treatment of disorders.

Collecting Clinical Data

It is difficult, but not impossible, for students to collect and analyze actual clinical data. Students trained in research methods and supervised by a skilled researcher can provide a meaningful service for harried health care providers who do not have an infrastructure to support research. Increasing requirements for accountability in health care service delivery have made it attractive for several local clinics, hospitals, and nonprofit organizations to develop joint projects with me and my students (see examples in Table 9.3). The organizations receive research resources, my students have the opportunity to apply their skills to real problems, and I get answers to interesting clinical questions. Part of what makes negotiations with clinical settings attractive is that the technical sophistication of the students' skills often exceeds those of the

TABLE 9.3

Research as Service: Health Care Research Needs and Complementary Student Research Skills

Health Care Research Needs	*Complementary Student Research Skills*
Project design	Psychosocial conceptual models
	Quasiexperimental research designs
	Observational studies
Data collection	Structured interviews of patients or staff
	Archival data retrieval
	Videotaping and behavioral coding
	Survey design and administration via Web
Data management	Relational database design
	Data entry
	Data format for import or export
Data analysis	Results
	Statistical analysis
	Interpretation
	Graphs and tables
Manuscript support	Literature searches
	Report drafts

professionals in the partnership, either because they never learned the statistics or their skills have atrophied from disuse.

Managing these studies requires flexibility and advance course planning because the form of the study dictates content of lectures, reading material, and training the students must have. Overestimates of the numbers of participants who might meet study eligibility, Institutional Review Board (IRB) protections, and naive optimism about prospective participants' willingness usually mean the clinical researcher should conservatively adjust projections of the host clinical setting.

In establishing the professional collaborations in advance, the most difficult issue is convincing health care professionals that undergraduate students can in fact understand the particulars of their research question and will be taught and supervised well enough to support a meaningful project. I often negotiate IRB approvals for class projects in advance of a semester's beginning. The demands of the professional setting dictate communicating aspects of professionalism to the students such as dress codes and maintaining their safety. Some medical settings require students to wear identification badges and complete volunteer orientation programs to both edify students and manage the host institution's liability.

Real-world projects also dictate important safeguards for the confidentiality of data collected in those settings. In my initial negotiations with host agencies I try to minimize the identifiers we need for patient data in order to better protect patients' confidentiality. I have prepared packets for educating students about APA ethical requirements for patient confidentiality and protection, and also a primer on the Health Insurance Portability and Accountability Act (HIPAA) requirements (see the CD accompanying this volume).

Methodological and Statistical Distinctiveness in the Course

Because our students must sample research methods courses from several areas, in this course I can focus on some of the contrasting elements of clinical research. For example, we discuss quasiexperimental methodologies and their strengths and limitations. Because clinical studies are frequently more confounded than other types of studies, I emphasize the use of statistical controls (e.g., analysis of covariance) and multivariate statistics (e.g., MANOVA, multiple regression) as strategies to design for and analyze the variables in a study. Students read articles representing each of the techniques as applied to a clinical problem, and we work example problems with the Statistical Package for the Social Sciences (SPSS). These problems include some type of clinical effect, and I expect students to use an appropriate statistical analysis to explain it.

Summary

My clinical research methods course is essentially a second course in both clinical psychology and statistics. Compared to most abnormal psychology courses, the class goes into more content detail about models of assessment and intervention. The class offers some advanced statistics as well. Given the substantial competition for places in APA-approved graduate programs in clinical psychology, the course is particularly good preparation for those candidates, as well as psychosocially oriented premedical undergraduates, important considerations for faculty and students at undergraduate institutions. The professional contacts and enrichments associated with negotiating community research projects also gratify faculty trained in scientist–practitioner models.

SIMULTANEOUSLY TEACHING METHODS AND SURVEY COURSES KRISTI S. MULTHAUP

For several years, I taught Psychology of Aging as a survey course. In a year when we had so many majors that we needed more seats in methods courses, I tried an experiment: I dropped the ceiling of my survey course by five so I could offer a five- or six-person lab section for an Adult Development methods course. The experiment was so successful that I have repeated the experience, and another colleague has followed suit with his Learning course.

Figure 9.1 shows how I structure the courses together. Students from the research methods course and students from the survey course attend the same lecture. I give tests to students in both courses on the same day, but they are different tests with more essays for the methods course students. Students in both courses do some common writing assignments. One of the major differences in the courses is that the students in the

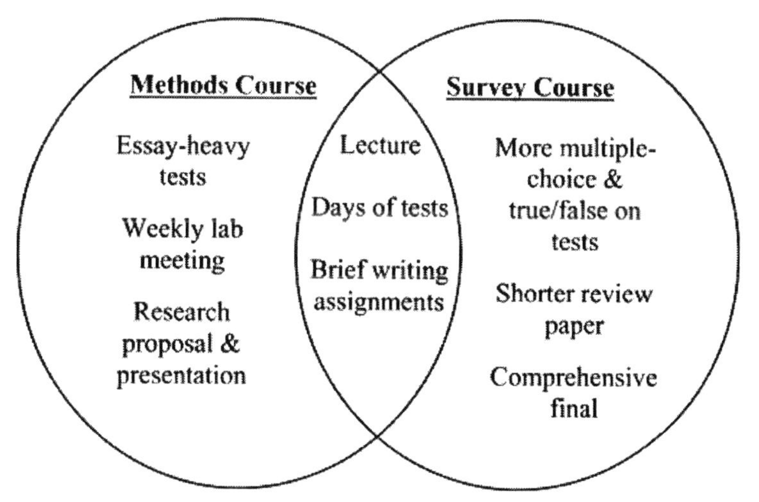

Figure 9.1. The intersection contains aspects of the course that are the same for the methods and survey courses; the other areas contain unique aspects of each course.

methods course have a lab meeting each week. That meeting has been 90 min, but I will be making it a 2-hr meeting in the future. The lab requires students to plan projects, collect data, check data entry, and analyze data. These activities culminate in lab reports. Also unique to the methods course is that those students bring all of their skills together at the end of the semester with a research proposal and a 25-min presentation to both classes, including 5 min for questions. The presentation occurs at least a week before the written proposal is due so that students can use the questions they receive and the feedback from their methods class peers to improve their proposals. By contrast, the survey course culminates in a shorter review paper and a comprehensive final exam.

The goal for the methods class is to give students experience with a variety of research methods used in the adult development literature. We do archival, experimental, and survey projects. I discuss each of these in turn, with a brief interlude about the first lab report before discussing the survey project.

Archival Project

We begin the archival project in the first week of the semester as we talk about stereotypes of older adults. We read McConatha, Schnell, and McKenna (1999) regarding how *Time* and *Newsweek* advertisements portray older adults. Students discuss how they could do a follow-up study, and I focus the discussion on how they can build on past work with a useful contribution. Examples of variables that students have chosen to explore further are selecting advertisements from a range of decades and including gender in the coding process. Coding schemes from prior research, such as McConatha et al.'s, form the basis of our scoring criteria (see companion CD to this volume). I pull the relevant criteria together and teach them to the students, emphasizing the need to be consistent when scoring. For each sampled issue of *Time* and *Newsweek*, two students independently score the advertisements, and then they meet to record their differences and work out any disagreements. After that, as a group we discuss issues of reliability and any problems with the scoring criteria. I then combine the data into one large data file in Excel®. I analyze the data in advance and develop a handout that guides the data analysis. Importantly, each student does each step of the data analysis at his or her own computer, although we progress through the analyses as a group, constantly talking about what the data mean.

Experimental Project

The experimental project has posed a larger challenge because we do not have time to collect data from older or middle-aged adults in the few weeks we have for this section. I have met this challenge in two ways: First, I had a data set from older adults awaiting the students. Each student went through the experiment as a participant to better understand the task and then the students scored the data. Students double-checked their work with a partner and I emphasized the need for such checking, particularly with hand-scored data. Second, I provided data already collected from older adults, and students collected data from college students for comparison. In this case I emphasized the need to use consistent instructions for all participants (we use a script for the experimental sessions) and the proper treatment of participants. In either case, upon data scoring or collection completion, I follow the same procedure as described in the archival project section.

At this point in the semester, students are prepared to write a lab report as an APA-style manuscript. They have the choice of writing about the archival project or the experimental project. We discuss this assignment several weeks before it is due so students have plenty of time to ask questions—that is particularly important for students who are writing their first lab report. In the discussion, I emphasize that students should tell a story about data and highlight articles we have read that are good models for telling stories. I also emphasize that APA style should not be copied from printed articles; students must use the APA manual for style advice. I have begun assigning Szuchman (2005) in my methods courses to help students with APA style (see Multhaup, 2005, for an explanation of how I use it).

Survey Project

During the several weeks that students have to write up their APA-style manuscript, we use lab time to prepare the survey project. We read and discuss background articles on two topics so there will be two parts to the survey we develop. As in the archival project, we discuss ideas for a follow-up project that can build on the prior research. Once students agree on a project, they work together to get materials developed, copied, and collated and to get the surveys out. As the data have come in, I often enter it and then use lab meetings for students to check the hand entry. For data analysis, I follow the same procedure as described in the archival project section. At the conclusion of the survey project, students are ready for their second lab report. Again they have a choice; they can choose either part of the survey for their report. However, this time students must tell their story in a limited space because the assignment is to do a conference-style poster in PowerPoint®.

Research Proposal and Presentation

After these three hands-on projects and lectures on a wide range of topics on aging, students draw everything together and develop a research proposal. In the 10th week of classes, students begin their search for a topic, which gives them 5 weeks to develop their presentation while we finish the survey project. I tell students to find an area that interests them, find out what has been done in that area, suggest a next step or variation on what has been done, and explain why it is important to do that step. I acknowledge that the assignment may be intimidating, and then I remind them that I do not expect them to revolutionize adult developmental research (although it is OK if they do!). I also remind them that we took small steps in the projects that we developed together and they need to use those as a model. This is also a natural point to talk again about crediting others for their ideas.

As explained earlier, students make class presentations at least a week before the proposals are due. For each presentation, methods students write at least one "applaud" and at least one constructive criticism. I type these comments so each student receives anonymous feedback. Students from the two courses get different things from the presentation experience. The methods students gain oral communication skills practice, get the chance to be experts in an area, and get feedback before their proposal is due. Students in the survey course get to hear about a variety of topics—some that were not covered in detail in the rest of the course—and see their peers doing advanced work. Students in the survey course have indicated that they enjoy the presentations. The presentations also offer both groups of students practice giving constructive feedback to their peers with questions and, for the methods students, their written comments.

GENERAL CONCLUSIONS

As the four course descriptions have illustrated, Davidson College psychology majors obtain broad and deep training in research methodology through hands-on laboratory experiences. They learn to read primary sources, design studies to investigate important questions, collect and analyze data, and disseminate results in professionally realistic ways. Their small-group collaborative work mirrors collaborative efforts by research scientists in many fields, not just psychology. Also, like some clinical and industrial/organizational research, some of our courses have a strong service component—the findings that students generate are immediately useful to an individual or organization that provided practical support for their research. From each of our courses, some students turn these experiences into undergraduate senior theses or other independent research projects, often based on the research proposal they generated at the end of a course. Their varied experiences across courses give them tools in their intellectual arsenals so they are in a better position to select appropriate methodology to investigate future research questions that interest them. Graduates report that because they have had so much practice doing research and writing reports, they are able to generalize these critical thinking skills to employment settings, as well as the more obvious extension to graduate school research demands. We therefore encourage faculty at other schools to insert hands-on laboratory experiences in their research methods courses and offer as many of those courses as possible.

REFERENCES

American Psychological Association. (2002). Ethical principles of psychologists and code of conduct. *American Psychologist, 57,* 1060–1073.

Bronfenbrenner, U. (1977). Toward an experimental ecology of human development. *American Psychologist, 32,* 513–531.

Gorenstein, E. E., & Comer, R. J. (2002). *Case studies in abnormal psychology.* New York: Worth.

Halgin, R. P. (Ed.). (2005). *Taking sides: Clashing views on controversial issues in abnormal psychology* (3rd ed.). Dubuque, IA: McGraw-Hill/Dushkin.

Halgin, R. P., & Whitbourne, S. K. (Eds.). (1998). *A casebook in abnormal psychology: From the files of experts.* New York: Oxford.

Jongsma, A. E. (2001). Therascribe: The Treatment Planning and Clinical Record Management System for Mental Health Professionals (Version 4.0) [Computer software]. New York: Wiley.

McConatha, J. T., Schnell, F., & McKenna, A. (1999). Description of older adults as depicted in magazine advertisements. *Psychological Reports, 85,* 1051–1056.

Multhaup, K. S. (2005). Review of the book *Writing with style: APA style made easy* (3rd ed). *The Journal of Undergraduate Neuroscience, 3,* R1–R2. Retrieved June 2, 2005, from http://www.funjournal.org/downloads/multhaup.pdf/

Murray, J. E. (1997). Flipping and spinning: Spatial transformation procedures in the identification of rotated natural objects. *Memory & Cognition, 25,* 96–105.

Rensink, R. A., O'Regan, K., & Clark, J. J. (1997). The need for attention to perceive changes in scenes. *Psychological Science, 8,* 368–373.

Rhodes, G., Jeffery, L., Watson, T. L., Clifford, C. W. G., & Nakayama, K. (2003). Fitting the mind to the world: Face adaptation and attractiveness aftereffects. *Psychological Science, 14,* 558–566.

Rhodes, G., Yoshikawa, S., Clark, A., Lee, K., McKay, R., & Akamatsu, S. (2001). Attractiveness of facial averageness and symmetry in non-Western cultures: In search of biologically based standards of beauty. *Perception, 30,* 611–625.

Shepard, R. N., & Metzler, J. (1971). Mental rotation of three-dimensional objects. *Science, 171,* 701–703.

Szuchman, L. T. (2005). *Writing with style: APA style made easy* (3rd ed.). Belmont, CA: Wadsworth.

Theeuwes, J., Kramer, A. F., Hahn, S., & Irwin, D. E. (1998). Our eyes do not always go where we want them to go: Capture of eyes by new objects. *Psychological Science, 9,* 379–385.

CD MATERIALS

Author – file contents – file type

Ault

Mission statement (pdf)

Barton

Clinical data sets folder

- Data Sets for Clinical Research methods (Word)
  - Read this for details concerning data sets

- Analysis of Covariance (Excel)
- MANOVA (Excel)
- Factor analysis (Excel)
- Multiple Regression (Excel)

Confidentiality Assurance (Word)
Project Management Forms (Word)
Useful URL (Word)

Multhaup

Archival scoring instructions (Word)
Archival data setup (Excel)

Munger

Mental rotation folder

- Readme with details (pdf)
- Stimuli (jpg files)
- PowerPoint outlining analysis
- Example data (Excel)

Facial attractiveness folder

- Readme with details (pdf)
- Stimuli (jpg files)
- Example data (Excel)

Chapter **10**

Innovative Approaches to Teaching Statistics and Research Methods: Just-in-Time Teaching, Interteaching, and Learning Communities

Kenneth E. Barron, James O. Benedict, Bryan K. Saville, Sherry L. Serdikoff, and Tracy E. Zinn
James Madison University

As a discipline, psychology includes a wide array of topics, and curriculum requirements vary considerably across different programs (Perlman & McCann, 1999). However, statistics and research methods courses are virtual constants in psychology curricula (Bailey, 2002). Therefore, it is not surprising that the teaching of these two courses is the focus of many articles published in journals such as *Teaching of Psychology*. Obviously, psychology instructors are interested in suggestions on how better to teach the material covered in statistics and research methods.

Considerable research over the past 30 years shows that lecture-based methods of instruction tend to be less effective than other alternative methods, and given the great deal of focus in the literature on how to teach statistics and research methods, this might especially be the case for these courses. Yet the majority of college teachers continue to follow the time-honored tradition of lecturing in their classrooms (Benjamin, 2002). There are likely many reasons for the continued use of lecture-based methods, and although we do not wish to speculate on all of them, some reasons might include: (a) faculty time is limited (Wright et al., 2004), and lectures are relatively easy to prepare; (b) many teachers simply adopt methods their teachers used; and (c) some administrators may not be supportive of teachers who take novel pedagogical approaches to teaching (e.g., Buskist, Cush, & DeGrandpre, 1991).

We feel fortunate to teach in the Department of Psychology at James Madison University (JMU), where effective teaching is valued, and the use of novel, nontraditional and innovative methods of classroom instruction is encouraged. In a related vein, JMU has also long been a leader in the area of assessment (e.g., Stoloff, Apple, Barron, Reis-Bergen, & Sundre, 2004). One result of this departmental interest in assessment is that faculty in the Department of Psychology continually rethink the best approaches to teaching their courses and evaluate the efficacy of any new teaching methods. Resulting from this focus are the three innovative methods of teaching statistics and research methods discussed here.

Although there are many ways to evaluate the effectiveness of these new methods of teaching statistics and research methods, we believe that Astin's (1993) model of student success is especially helpful in this case. In *What Matters in College?: Four Critical Years Revisited*, Astin described the results of an extensive longitudinal study in which he collected data from nearly 25,000 college students at over 300 higher education institutions. From these data, Astin developed a model of student success that contains three important variables: (a) *inputs*, which are "those personal qualities [such as personality, self-predictions of academic success, and parental income] the student brings initially to the education program" (p. 18); (b) *environment*, which refers to "the student's actual experiences during the educational program" (p. 18) and includes specific classroom and extracurricular experiences; and (c) *outputs*, which are "the 'talents' [that students] develop in our educational program" (p. 18) and include grade point average, degree completion, and overall satisfaction with college, to name a few. In other words, Astin's model suggests that students' personal qualities and college experiences work both individually and collectively to affect various markers of student success.

Of the environmental variables—those variables over which college instructors have the most influence—Astin believed that three were particularly important with regard to student learning: student–teacher interaction, student–student interaction, and time on task. Importantly, college instructors have the opportunity to control these three variables by carefully planning the structures of their courses. The methods described here provide three models for how instructors can design their statistics and research methods courses in order to focus on these important environmental variables. Therefore, the purpose of this chapter is twofold: (a) to describe Just-in-Time Teaching, Interteaching, and Learning Communities—innovative approaches to classroom instruction that we use to teach statistics and research methods at JMU; and (b) to describe how each of these approaches positively affects student–teacher interactions, student–student interactions, and time on task. By developing statistics and research methods courses that focus on enhancing these three variables, we are more likely to positively impact our students' learning.

JUST-IN-TIME TEACHING

Just-in-Time Teaching (JiTT) is a relatively new pedagogy that uses the Internet to prepare students and teachers for highly productive classroom experiences. Novak, Patterson, Gavrin, and Christian (1999) created JiTT to improve the teaching of college physics. However, JiTT has several objectives that instructors can apply to teaching many courses in the psychology curriculum, especially statistics and research methods:

1. JiTT encourages students to attempt to answer questions that will form the basis for subsequent class discussion, without taking up additional class time. The teacher places the questions on a Web form, and students submit their answers just before class. The questions might require students to solve certain statistical problems or to analyze research case studies.
2. JiTT provides students with near-immediate feedback for their attempts at solving problems.
3. JiTT provides teachers with valuable information about what content they should cover during class. For example, if a teacher asks her students to differentiate and give examples of Type I and Type II errors, she can quickly determine by looking at the submitted responses if she needs to discuss these concepts in class. Thus, JiTT can serve as an important tool for formative assessments of student learning in conjunction with more traditional summative assessments.
4. JiTT requires that teachers discuss some of the student submissions, which provides feedback to students and serves as the basis for class discussions.

Implementation of JiTT

At JMU, a number of faculty members have adopted JiTT techniques in their psychology classes. A teacher uses JiTT in the following way. First, the teacher identifies two or three objectives that reflect the focus of the day's lecture and then writes a question for each objective. Typically, each question is difficult enough to elicit common student errors and misconceptions. For example, in a statistics class, a typical JiTT question focusing on the concept of *variability* might ask students to compare the F ratio with the signal-to-noise ratio. Second, the teacher posts these questions 2 or 3 days before class on the Internet using a Web-based form. At JMU, users of JiTT refer to these questions as PreClassQuestions (PCQs); others have called them WarmUps or PreFlight Questions (e.g., Novak et al., 1999). Then, during this 2–3 day period, students read the appropriate material and submit their answers to the questions. Students can work alone, or the teacher can ask students to interact with another student and submit their responses as a team. The teacher usually includes on the Web form a "comment box" that allows students to ask general questions or make comments about the course.

Novak et al. (1999) called this approach Just-in-Time Teaching because students must submit their answers just before class. For example, students in the second author's class must turn in their answers at least 2 hr before class starts but may submit their answers a day or more in advance. Consequently, the teacher has time to look over students' responses before class; it also helps the teacher (a) determine if students understand the content and (b) identify which submissions should serve as the basis for class discussion. Thus, one of the immediate benefits of using JiTT is that teachers can better diagnose what concepts they should cover and discuss in class. Without an assessment tool like JiTT, a teacher may waste valuable class time teaching concepts that students already know and understand. To provide students with feedback, the teacher downloads two types of answers: those that demonstrate typical confusions or misconceptions and those that demonstrate excellent understanding of course material. The teacher displays these anonymous examples, which then serve as the basis for a subsequent lecture. Although only 5% to 10% of the final grade is based on the submissions, the second author has found that his students reliably submit responses to the posted PCQs each week. Some instructors give credit for submissions, independent of the quality of answer; the second author has found that students spend more time on their submissions if part of the credit is dependent on the correctness of the answers.

To date, the second author has used JiTT while teaching statistics, research methods, general psychology, and a course in advanced statistical procedures. Typically the Web form includes two short-answer or essay questions, a difficult multiple-choice question, and a comment box. The Web form could use all multiple-choice questions, particularly in a very large class, or all essay questions in a small class. In terms of the number of questions included, Novak et al. (1999) and others have used only three questions on each assignment, which seems to fit well with the number of objectives for each class period. And although Novak et al. reported that they use JiTT for each class period, the second author typically uses JiTT once per week.

To determine the success of this approach, Benedict and Anderton (2004) used two types of evaluations. As a cognitive evaluation, they compared students' final exam scores in a JiTT-based statistics class to scores in a similar, but non-JiTT-based statistics class and found that students in the JiTT-based course scored significantly higher. As an affective evaluation, they asked students to react to several questions measuring their attitudes about the use of JiTT. Benedict and Anderton found that more than two thirds of students in the JiTT-based course thought that the approach was extremely effective for learning statistics and improving their problem-solving skills. These results are similar to those obtained by the second author in other courses.

Astin's (1993) Model

With regard to Astin's (1993) model, JiTT increases student–teacher interaction, student–student interaction, and time on task. However, the main focus of JiTT is to increase student-teacher interaction, which benefits both students and teachers. First, JiTT benefits students by providing a new channel of communication to teachers. Although students normally convey their knowledge to teachers on quizzes and exams, PCQs encourage more of a tutorial relationship with teachers. For example, a student may submit a comment halfway though the statistics course stating that he or she does not know whether "$p < .05$" or "$p > .05$" means that a test is statistically significant. Typically, a student might not admit this concern in class but may do so in the comment box of the JiTT form, which may prompt the teacher to spend some additional class time explaining decision rules. Because JiTT requires students to convey their understanding of concepts to their teachers just before class, the teacher can lead a discussion based on sample responses with the aim of creating the "perfect answer," which provides useful feedback to students. Thus, JiTT not only facilitates understanding of content but also helps to reduce possible anxieties and to personalize the learning experience for students. It simulates a one-to-one tutorial experience for students.

Second, JiTT may also increase student—student interaction. JiTT certainly increases such interaction when students prepare and submit responses as teams. Students also learn from each other when they have a chance to see their peers' answers in class. The presentation of student writing during class allows a unique learning experience because in most courses, students seldom have the chance to read each other's writing. When Benedict and Anderton (2004) asked students in their psychological statistics classes how they felt about the instructor displaying and discussing anonymous student answers, they found that 90% reported it to be helpful. Thus, students find it beneficial to see their own and others' attempts at solutions

Finally, JiTT also increases the amount of time that students spend on task (i.e., learning course content). For example, in the past, the second author used to spend time in his research methods course engaging in group discussion of case studies, identifying specific research designs, or labeling threats to internal validity. With JiTT, students now come to class already having spent time thinking about and classifying the case studies. Each student has spent additional time outside of class covering this material. (These preclass submissions also give the teacher 20 to 30 min more class time.) With regard to how much time students spend on their submissions, the second author has found that students in his statistics course reported spending an average of an additional 45 min (*SD* = 28.74) composing their solutions to the three questions that require application of some complex procedural knowledge. As such, using JiTT three times a week, as Novak et al. (1999) did, would likely increase time-on-task considerably.

One minor drawback to using JiTT, however, is that it requires more preparation time for teachers. For example, it can take several hours to learn how to create Web forms; and, initially, it may take 1 to 2 hr each week to create the PCQs and an additional 1 to 2 hr to read and grade the submissions for a class of 35 students. After using the technique for several years, the second author now spends an extra 30 to 60 min each week, and the technique is still received favorably by students and by the instructor. Because it increases student–faculty and student-student interactions, increases time on task, makes class time more productive, and has negligible drawbacks, JiTT is a pedagogical approach worthy of use in both statistics and research methods courses.

INTERTEACHING

Behavior-analytic approaches to classroom instruction, including Programmed Learning (e.g., Holland & Skinner, 1961), Precision Teaching (e.g., Lindsley, 1964), and Direct Instruction (e.g., Becker & Engelmann, 1978), tend to be superior to other, more traditional methods of classroom instruction (e.g., Moran & Malott, 2004). The most well-known and thoroughly studied of these approaches is Keller's (1968) Personalized System of Instruction (PSI), which features (a) self-pacing, (b) an emphasis on writing, (c) infrequent motivational lectures, (d) a mastery criterion, and (e) the use of student proctors. After Keller (1968) published his seminal paper, PSI quickly became a popular alternative to standard methods of classroom instruction—within the next decade it had established itself empirically as superior to several other instructional methods (for a review, see Buskist et al., 1991). More recently, however, the use of PSI and other behavior-analytic methods of classroom instruction has become less common (see Fox, 2004), despite their proven effectiveness (e.g., Binder & Watkins, 1990; J. A. Kulik, C. C. Kulik, & Cohen, 1979).

There are several possible reasons why instructors discontinued use of these methods, even though they tend to be more effective than traditional methods (see Buskist et al., 1991; Fox, 2004; Saville, Zinn, Neef, Van Norman, & Ferreri, in press). First, instructors are often reluctant to adopt unique methods, especially when they deviate from more common methods of instruction (Buskist et al., 1991). Second, these methods often require extensive preparation and resources (e.g., Boyce & Hineline, 2002; Lindsley, 1992), which may discourage some instructors from adopting them. Finally, general misunderstandings of the principles that underlie these methods have led some researchers to discount their efficacy (e.g., Lindsley, 1992) or apply the methods incorrectly.

In response to such resistance, Boyce and Hineline (2002) developed *interteaching*, a new method of classroom instruction based on experimentally derived principles of behavior. They suggested that interteaching "retains some key characteristics of Keller's personalized system of instruction and precision teaching, but offers greater flexibility for strategies that are based on behavioral principles" (p. 215). It also attempts to resolve some of the problems with other behavior-analytic methods of instruction (see Saville, Zinn, & Elliott, 2005).

Interteaching: General Procedure

A typical interteaching session proceeds as follows (see Boyce & Hineline, 2002; Saville, Zinn, Neef et al., in press, for further details). First, the instructor constructs a preparation guide ("prep guide") that students use to guide themselves through the assigned readings. Although the content on the prep guides varies depending on the material to be covered, it typically contains a variety of items, with questions ranging from definitional and comprehension to analytical and evaluation (see Bloom, 1956). For example, in statistics, one question might ask students to define probability and relate probability to the normal curve, focusing on definitions and computation. Another question might ask students to compare the persuasiveness of an argument based on statistical probability to one based on an anecdotal story and analyze under what conditions statistics might be more or less persuasive.

To ensure that students have ample time to work through the prep guide items before class, we ordinarily make the prep guides available ordinarily to students at least a few days before a given interteaching session. For a typical statistics course that meets three times per week, we would post prep guides on a course Web site (e.g., Blackboard, WebCT) the week before they are due. The prep guides would also typically cover 10 to15 pages of text, depending on the difficulty of the material.

During the interteaching session, students work in dyads or sometimes triads and discuss the items contained on the prep guide. The instructor moves about the room while students are engaged in discussion to answer any questions that students might have and to facilitate student discussion of the material. Although the length of the discussions may vary, they typically account for approximately 75% of the assigned class time (e.g., 55 min in a 75-min class).

After they finish discussing the prep guide items, students fill out an "interteaching record," on which they list the following information: (a) name of their partner; (b) how long their discussion lasted and if they had enough time to finish; (c) quality of their discussion (rated on a 7-point Likert scale), and what factors contributed to the quality (e.g., both people were prepared); (d) what items were difficult to answer and why; (e) any topics they would like the instructor to review; and (f) any other information that might be useful to the instructor. Based on this information, the instructor constructs a short lecture that begins the next class period. Typically, the lecture includes information designed to clarify items that students found difficult or confusing, as well as any supplemental information that might be relevant to the topic in question. Instructors know the material that students found difficult and thus can focus their lectures on only the material that students specifically requested. Consequently, "if the lecture is properly developed, its potency as a reinforcer is virtually assured" (Boyce & Hineline, 2002, p. 222).

Interteaching: Additional Components

There are two other components of interteaching that likely contribute to its efficacy—although additional research is needed to validate this statement. First, exam items come from material contained on the prep guides. In fact, Boyce and Hineline (2002) suggested that exams should contain at least two essay-type questions that come directly from the prep guides, as well additional objective questions (e.g., multiple-choice, fill-in-the-blank) pertaining to other material on the prep guides. As a result, students understand that there is a direct relation between material on the prep guides and the questions found on the exams, which is likely to increase preexam studying and reduce apprehension that sometimes comes with exam preparation (e.g., "I just don't know what to study."). Second, to increase the likelihood that students spend their time engaging in quality discussion during each interteaching session, Boyce and Hineline suggested the use of *quality points*. Specifically, students' exam scores are partially dependent on the performance of their partners: If students and their partners both do well on the essay questions on each exam, they receive additional points toward their course grades. Although quality points account for only a small percentage of each student's grade (e.g., 10% of the overall course grade), the inclusion of this cooperative contingency likely motivates students to teach one another as effectively as possible.

Is Interteaching Effective?

Although Boyce and Hineline (2002) provided rationale for why interteaching should be a useful alternative to other forms of classroom instruction, they provided only anecdotal evidence regarding its efficacy. Consequently, they urged others to collect experimental data to validate their claims. In a recent study, Saville, Zinn, and Elliott (2005) examined interteaching relative to two other forms of instruction: lecture and reading. They randomly assigned participants to one of four conditions—interteaching, lecture, reading, or control—and examined performance on a short, multiple-choice quiz. They found that participants in the interteaching group performed significantly better than participants in the other groups. However, because they conducted their study in a highly controlled laboratory setting, it is difficult to determine if these results would generalize to a traditional college course.

To extend Saville et al.'s (2005) results, Serdikoff (2004) compared interteaching to lecture in two sections of an undergraduate research methods course. During one summer session, 18 students heard lectures over the course material; the following summer, 18 students covered the same material with interteaching. Although the exams in the two courses varied slightly, most of the multiple-choice items were identical. Serdikoff observed that students in the interteaching condition performed considerably better on these items than students in the lecture condition. Because Serdikoff was unable to assign participants to interteaching or lecture randomly, however, it is possible that differences between the two groups existed prior to the introduction of the two instructional methods.

In a subsequent study, Saville, Zinn, Neef et al. (in press, Experiment 2) also examined interteaching and lecture in two sections of an undergraduate research methods course. Although they did not assign students to the two conditions randomly, the researchers attempted to control for any preexisting differences between groups (and other potential confounds) by alternating interteaching and lecture throughout the semester and by counterbalancing across sections. They observed that students (a) performed better on exams following interteaching, (b) performed better on a cumulative final exam following interteaching, (c) reported that they learned more with interteaching, and (d) reported that they enjoyed interteaching more than lecture.

Why is Interteaching Effective?

Although future research will help to elucidate exactly why interteaching enhances student learning, Astin's (1993) model of student success provides one possible explanatory framework. Like JiTT, interteaching may be effective because it tends to produce increases in student–teacher interaction, student–student interaction, and time on task. First, during interteaching sessions, the teacher's role is to move among groups, answer questions, clarify information, and facilitate discussion. These frequent interactions provide teachers with ample opportunities to learn more about their students, provide positive feedback, and increase student–teacher rapport. Second, interteaching requires students to work with one another to respond to material in the prep guides. Thus, students spend the majority of each interteaching class period interacting with other classmates and helping each other learn the material. Finally, because students must respond to prep guide items prior to coming to class each day, they necessarily spend more time engaging the course material.

Thus, interteaching seems to be an effective method for teaching statistics and research methods. As with any innovative method of teaching, however, instructors should be aware that there are certain caveats. First, interteaching is initially time-consuming. Although it may ultimately reduce the amount of time that teachers spend preparing course materials, it does take a considerable amount of time to construct good prep guides. In addition, anecdotal evidence (e.g., Boyce & Hineline, 2002) suggests that interteaching may not work as well if classes meet infrequently. Finally, students may initially be wary of interteaching for several reasons: Some may not like the amount of work that interteaching typically requires, and others express concern over the fact that part of their course grade is dependent on other students' performances (i.e., the quality points). However, we believe that the positive aspects of interteaching outweigh the negative aspects. If teachers wish to maximize students' learning, interteaching provides an effective and enjoyable alternative to other methods of classroom instruction. More often than not, students realize these positive qualities of interteaching as well.

CREATING A LEARNING COMMUNITY FOR STATISTICS AND RESEARCH METHODS

Another approach to improving student engagement and academic success in their coursework is the formation of learning communities. Although not a new concept in higher education, learning community initiatives are growing in popularity as a powerful intervention to increase student involvement in curricular and co-curricular activities (Gamson, 2000; Shapiro & Levine, 1999; Taylor, Moore, MacGregor, & Lindblad, 2003). A *learning community* is a group of students who are enrolled in a common set of courses, typically organized around certain themes. A learning community can also involve a residential living component, in which students live together while taking common courses together. Interested readers can find an excellent introduction about learning communities, as well as a clearinghouse of resources, at the National Learning Communities Project Web site http://learningcommons.evergreen.edu

The Development of a Residential Learning Community in Psychology at JMU

In 2000, JMU initiated a residential learning community program for first-year students. Each learning community is centered on a particular theme and set of coursework. For example, JMU currently has communities focused on particular majors (e.g. biology), special programs (e.g., honors courses), and general education coursework. In 2002, we were approached to develop a residential learning community for psychology. We debated at great length on what theme and coursework our community should have. Interestingly, the one course that seemed like the most probable candidate to include (introductory psychology) determined the ultimate direction for our community. With the growing number of students earning Advanced Placement (AP) credit for psychology, we realized that our applicants already may have already earned college credit for the course. Therefore, we decided to center our community on statistics and research methods, the next courses required in our major. Students often regard these courses to be the most challenging courses in our curriculum, and the learning community environment would provide a forum for additional interventions to help facilitate students' involvement with the material. For example, because students would be living together in the same residence hall, they could easily form study groups and seek each other out for additional assistance.

Additional advantages of creating a learning community were rethinking the actual coursework that we offer in our major and experimenting within this smaller group to determine if there is a better way to deliver our courses. Currently, JMU psychology majors must complete a two-course statistics and methods sequence. The first is Psyc 210 (Psychological Statistics), which introduces students to the different statistical tools used in psychology. The second is Psyc 211 (Psychological Research Methods), which introduces students to different research methods and how best to design and carry out research. Students take these courses in a specific order (Psyc 210 is a prerequisite for Psyc 211); in Psyc 210, students focus on learning statistical skills that they will need to apply in Psyc 211, when they engage in research.

Although a similar two-course sequence is found in most undergraduate psychology programs around the country, a common critique of this approach is that students often lack the necessary context to appreciate why they are learning about different tools in their statistics course and why knowing how to use these tools will be beneficial to them in psychology. One analogy that quickly captures what can happen when a statistics course is made a prerequisite for a research methods course is putting the cart before the horse. In other words, students learn specific tools used to answer questions in research before being taught what research is and why research is such an integral part of psychology. Thus, it may not be surprising that reteaching important statistical concepts is often necessary for students to complete projects in their research methods course and that there is a concern that students are not taking full advantage of what they can learn in their statistics course. Therefore, we decided to experiment with three new courses in our learning community that we labeled as Psyc 200A, 200B, and 200C.

Instead of keeping the content of these two courses separate (where students learn *statistical* concepts one semester and *research methods* concepts the next), we proposed Psyc200A and 200B (Psychological Research Methods and Data Analysis I & II) as a new two-course sequence that would integrate themes currently taught in Psyc 210 and Psyc 211. In these new courses, students shift in and out of units on research methods and statistics during each semester. First, we introduce students to a particular research design and what questions you can answer with it. Second, students learn about tools that they will need to analyze and draw conclusions using that design. Third, students engage in a hands-on research project to practice adopting that research approach and using the statistical tools associated with that research design. Our goal is to provide better context about why different methodological approaches and statistical tools are necessary for our field and necessary for us to be better researchers.

We created Psyc 200C (Orientation to Psychology and the Major) as a companion course to expose students further to the diversity of research topics within psychology, the different career paths students could pursue with training in psychology, and the unique opportunities that JMU psychology majors can experience. Over the course of the semester, students interact one-on-one with different psychology faculty in a weekly symposium format. Our typical format involves guest speakers facilitating a discussion on a particular area and profession within psychology (e.g., industrial/organizational psychology), providing examples of research in that area, and suggesting how students could get involved in research in that area. Thus, we try to use this course to reinforce continually the value of the research methods training that students are learning in Psyc 200A and B.

How a Learning Community Promotes Astin's Model of What Matters in College

Astin (1985; see also Astin, 1993) was quick to note how learning communities can play a pivotal role in increasing student involvement and success. First, participation in a learning community greatly increases student--student interaction. Our students took three classes together over the course of their freshmen year, which necessarily resulted in students interacting with each other on an ongoing basis. We also created regular assignments that required students to interact with each other outside of class. Second, participation in a learning community facilitates student–faculty interaction. For example, faculty teaching in the learning community had the chance to interact with students up to 6 hr a week in just regularly scheduled class time alone. We also designed Psyc 200C to promote student–teacher involvement with other psychology faculty throughout our department and to introduce students to our faculty's research (e.g., in this past year alone, our students met with 23 different psychology faculty at our university). Finally, residential learning communities greatly encourage time on task. Class discussions naturally spill over and continue in residence halls, and our students often report that being able to find a classmate at almost any time of day to discuss class assignments is one of the greatest benefits of the community. We also take advantage of holding regular office hours and review sessions at night in the residence hall.

Assessment of our Learning Community Initiative

Assessment evidence for the general benefits of a learning community to promote students' academic success is documented in a number of sources (see Taylor et al., 2003, for a review). Regarding our community, JMU's psychology department has an extensive, multidimensional assessment of graduating seniors (see Stoloff et al., 2004, for a review), including content knowledge of psychology (like statistics and research methods), writing, experiences in and out of the major, and growth along the 10 learning goals identified by the American Psychological Association (Halonen et al., 2002). As of 2005, we have completed our third year of the learning community program; therefore, our first cohort of students will be graduating next year and participating in an extensive senior assessment. However, we have tracked other markers while we wait for senior assessment data. For example, only 1% of nonlearning community freshmen psychology majors participated in independent research during this past year. In contrast, 40% of freshmen in the learning communities completed some form of independent research.

Although learning communities can greatly capitalize on Astin's (1993) model of student success, there are important cautions to note before embarking on this undertaking (see also Shapiro & Levine, 1999). First, it is time consuming for faculty due to the increased expectations for and level of student involvement, as well as additional administrative work. Second, there also were several challenges and unexpected findings that we encountered along the way. For example, we went into the Learning Community initiative naively expecting that students would all be intrinsically motivated and would gladly take advantage of all of the experiences that we would provide; this did not end up being the case and we were often frustrated by some students' lack of motivation. Also, although we expected and hoped that students' academic conversations would naturally continue in the residential hall, we were not prepared for the opposite spillover. For instance, roommate and hall conflicts frequently spilled into the classroom environment in negative ways. Finally, cliques tended to form within the community if we did not actively develop group dynamics and team building. However, appreciation and attention to each of these details in our subsequent years of offering our community has led to significant improvements in the quality of our experiences. In fact, the third year of the Learning Community was the most successful yet, and we anticipate the outcomes improving further next year.

FINAL THOUGHTS

Teaching statistics and research methods courses effectively can be difficult, time-consuming, and often frustrating. However, based on data we have collected, we believe that the approaches discussed in this chapter can be helpful in structuring a more effective learning environment, thus positively affecting both students and instructors. Each of these methods seems to have a positive impact on student–student interactions, student–teacher interactions, and time on task—three variables that Astin (1993) identified as important for student learning. However, the methods we describe are by no means the only ways of positively affecting these variables. Thus, we encourage teachers of psychology to embrace nontraditional and innovative methods of instruction, including those discussed herein, for teaching research methods and statistics. Although each method discussed here has required significant work to implement, the outcomes have been well worth it. We urge teachers to consider models of student success like Astin's—and ours—when constructing those initiatives.

AUTHORS' NOTES AND ACKNOWLEDGMENTS

Authorship order was determined alphabetically. The authors wish to thank Bill Buskist for organizing the symposium on which the present chapter was based and for providing valuable feedback on an earlier version of this chapter.

For further information on (a) Just-in-Time Teaching, contact Jim Benedict (benedjo@jmu.edu); (b) interteaching, contact Bryan Saville (savillbk@jmu.edu), Sherry Serdikoff (serdiksl@jmu.edu), or Tracy Zinn (zinnte@jmu.edu); or (c) learning communities, contact Kenn Barron (barronke@jmu.edu). All authors can be reached at Department of Psychology, MSC 7401, James Madison University, Harrisonburg, VA 22807.

REFERENCES

Astin, A. W. (1985). *Achieving educational excellence*. San Francisco: Jossey-Bass.

Astin, A. W. (1993). *What matters in college?: Four critical years revisited*. San Francisco: Jossey-Bass.

Bailey, S. A. (2002). Teaching statistics and research methods. In S. F. Davis & W. Buskist (Eds.), *The teaching of psychology: Essays in honor of Wilbert J. McKeachie and Charles L. Brewer* (pp. 369–377). Mahwah, NJ: Lawrence Erlbaum Associates

Becker, W. C., & Engelmann, S. (1978). Systems for basic instruction: Theory and applications. In A. C. Catania & T. A. Brigham (Eds.), *Handbook of applied behavior analysis: Social and instructional processes* (pp. 325–377). New York: Irvington.

Benedict, J. O., & Anderton, J. B. (2004). Applying the Just-in-Time Teaching approach to teaching statistics. *Teaching of Psychology*, 31, 197–199.

Benjamin, L. T., Jr. (2002). Lecturing. In S. F. Davis & W. Buskist (Eds.), *The teaching of psychology: Essays in honor of Wilbert J. McKeachie and Charles L. Brewer* (pp. 57–67). Mahwah, NJ: Lawrence Erlbaum Associates.

Binder, C., & Watkins, C. L. (1990). Precision Teaching and Direct Instruction: Measurably superior instructional technology in schools. *Performance Improvement Quarterly*, 3, 74–96.

Bloom, B. S. (Ed.). (1956). *Taxonomy of educational objectives, Vol. 1: Cognitive domain*. New York: Longmans Green.

Boyce, T. E., & Hineline, P. N. (2002). Interteaching: A strategy for enhancing the user-friendliness of behavioral arrangements in the college classroom. *The Behavior Analyst*, 25, 215–226.

Buskist, W., Cush, D., & DeGrandpre, R. J. (1991). The life and times of PSI. *Journal of Behavioral Education*, 1, 215–234.

Fox, E. J. (2004). The Personalized System of Instruction: A flexible and effective approach to mastery learning. In D. J. Moran & R. W. Malott (Eds.), *Evidence-based educational methods: Advances from the behavioral sciences* (pp. 201–221). New York: Academic Press.

Gamson, Z. F. (2000). The origins of contemporary learning communities: Residential colleges, experimental colleges, and living-learning communities. In D. Dezure (Ed.), *Learning from change* (pp. 113–116). Washington, DC and Sterling, VA: American Association for Higher Education and Stylus Publishing.

Halonen, J. S., Appleby, D. C., Brewer, C. L., Buskist, W., Gillem, A. R., Halpern, D., et al. (Eds.). (2002). *Undergraduate major learning goals and outcomes: A report*. Washington, DC: American Psychological Association. Retrieved May 1, 2005, from http://www.apa.org/ed/pcue/taskforcereport2.pdf

Holland, J. G., & Skinner, B. F. (1961). *The analysis of behavior*. New York: McGraw-Hill.

Keller, F. S. (1968). Good-bye teacher... *Journal of Applied Behavior Analysis*, 1, 79–89.

Kulik, J. A., Kulik, C. C., & Cohen, P. A. (1979). A meta-analysis of outcome studies of Keller's Personalized System of Instruction. *American Psychologist*, 34, 307–318.

Lindsley, O. R. (1964). Direct measurement and prosthesis of retarded behavior. *Journal of* Education, 147, 62–81.

Lindsley, O. R. (1992). Why aren't effective teaching tools widely adopted? *Journal of Applied Behavior Analysis*, 25, 21–26.

Moran, D. J., & Malott, R. W. (Eds.). (2004). *Evidence-based educational methods: Advances from the behavioral sciences*. New York: Academic Press.

Novak, G. M., Patterson, E. T., Gavrin, A. D., & Christian, W. (1999).: *Just-in-Time Teaching Blending active learning with Web technology*. Upper Saddle River, NJ: Prentice-Hall.

Perlman, B., & McCann, L. I. (1999). The structure of the psychology undergraduate curriculum. *Teaching of Psychology*, 26, 171–176.

Saville, B. K., Zinn, T. E., & Elliott, M. P. (2005). Interteaching vs. traditional methods of instruction: A preliminary analysis. *Teaching of Psychology*, 32, 161–163.

Saville, B. K., Zinn, T. E., Neef, N. A., Van Norman, R., & Ferreri, S. J. (2006). A comparison of interteaching and lecture in the college classroom. *Journal of Applied Behavior Analysis, 39*, 49–61.

Serdikoff, S. L. (2004). [Alternative techniques for teaching psychological research methods: A comparison of lecture and interteaching]. Unpublished raw data.

Shapiro, N. S., & Levine, J. H. (1999). *Creating learning communities: A practical guide to winning support, organizing for change, and implementing programs*. San Francisco: Jossey-Bass.

Stoloff, M., Apple, K., Barron, K. E., Reis-Bergen, M. J., & Sundre, D. A. (2004). Seven goals for effective program assessment. In D. Dunn, C. Mehrotra, & J. Halonen (Eds.), *Measuring up: Assessment challenges and practices for psychology* (pp. 29–46). Washington, DC: American Psychological Association.

Taylor, K., Moore, W. S., MacGregor, J., & Lindblad, J. (2003). *Learning community research and assessment: What we know now*. National Learning Communities Project Monograph Series. Olympia, WA: The Evergreen State College, Washington Center for Improving the Quality of Undergraduate Education in cooperation with the American Association for Higher Education.

Wright, M. C., Assar, N., Kain, E. L., Kramer, L., Howery, C. B., McKinney, K., et al. (2004). Greedy institutions: The importance of institutional context for teaching in higher education. *Teaching Sociology, 32*, 144–159.

Chapter **11**

Teaching Ethics in Research Methods Classes

Edward P. Kardas
Southern Arkansas University
Chris Spatz
Hendrix College

This chapter addresses an approach to teaching ethics in research methods classes. Practices that are ascribed to "our classes" are those that one or both of us use and that we both endorse. Our two institutions are a regional public university (Kardas) and a small liberal arts college (Spatz), both in Arkansas. In this chapter, we discuss ethics in an increasingly specific way: ethics, ethics in science, and ethics in psychology. We divide the researcher's ethical responsibilities into two categories: responsibilities to participants and responsibilities to psychology itself. We end the chapter with a discussion of ethics in animal research, an example of a Socratic ethical debate used in class, and a technique for ensuring ethical student research.

Ethics and research methods are two inseparable matters; ethics is not just another methods course topic. Instead, ethics is part of the very fabric of research itself. Consequently, we teach ethics throughout our stand-alone research methods courses, which, in both our curricula, follow an introductory statistics course. We start early with historical examples, later cover the *Ethical Principles of Psychologists and Code of Conduct* (American Psychological Association, 2002), engage students in Socratic debates of ethical scenarios, and finally, insist on proper research ethics in class research.

We require original student research as part of our methods courses. Students design and defend a research prospectus on both its ethical and methodological components. Only then may they conduct their research. In brief, we believe in the importance of interweaving ethics and method into a seamless whole.

HISTORY OF ETHICS

Our coverage of ethics highlights the larger study of ethics, which is an underpinning of moral behavior and also its own field of study. Our individual approaches in the classroom vary, but we teach psychology as a discipline embedded in a historical matrix. Because most of our students are from Arkansas and nearby states, most are familiar with a Christian religious ethical code. As a result, our students arrive with a fairly sophisticated, but often narrow, conception of moral behavior. Their information base, prejudices, and worldviews both facilitate and complicate our tasks as teachers. Teaching research ethics as a religion—free ethical code is easy, but confronting long-held beliefs such the sinfulness of homosexuality is not. Eventually, we show that science's ethical code evolved independently of religiously based moral codes. By the end of the semester, we find that most students learn and use research ethics without believing that they are compromising their personal ethical beliefs. However, we are vigilant against teaching anything other than scientific research ethics.

We discuss briefly the origin of religiously based codes based on the *Torah*, the *Bible*, the *Koran*, the *Analects of Confucius*, the *Book of Mormon*, and the teachings of Buddha. We are not attempting to teach comparative religion. Instead, we show how religiously based moral codes originated in many areas of the world and that some possess enormous staying power. Furthermore, many of these moral codes evolved into present-day legal codes, which constitute additional evidence of their pervasiveness. We note that the study of ethics is an independent academic subject, too. The teaching point is to show that ethics is a longstanding and respected academic topic. Furthermore, we argue that the study of ethics does not need to be necessarily linked to religious formulations.

We also distinguish between absolutist and relativistic thinking and show how those forms of thinking affect the study and practice of ethics. We define *ethical absolutist thinking* as moral decisions that are made on the basis of tradition or codified text. We define *ethical relativistic thinking* as moral decisions that vary based on norms that are local in place or time. Current controversial issues often provide a forum for class discussion of absolutist versus relativistic thinking. In class, we have used abortion rights, the teaching of evolution, and the causes of homosexuality as vehicles for such discussions. For some of our students, especially those high in religiosity, such discussions have caused some unease or discomfort because they believe we are challenging their personal beliefs.

ETHICS IN SCIENCE

Walter Reed's research on yellow fever in Cuba during the early years of the 20th century is a noteworthy place to begin discussion of scientific ethics because it is one of the first instances in which a researcher used a consent form (Pierce, 2003). Reed did so because previous research on yellow fever resulted in the unexpected deaths of research participants (including a member of his team, Jesse Lazear). Because yellow fever does not affect animals, humans had to be the participants. The research team began their investigations by first infecting themselves with yellow fever by having a mosquito bite an infected person and 2 weeks later, having the same mosquito bite them. When they continued their project, which included many more volunteer participants, each had to sign the consent form. The research results clearly implicated mosquitoes as the vector of yellow fever. Controlling mosquito populations became the method of combating yellow fever and those efforts proved extremely successful in Cuba and elsewhere.

Beginning our discussion of scientific ethics using a positive example of participant consent is important to us. Illustrating negative examples of participant consent is, unfortunately, far too easy and starts students' studies of ethics on the wrong foot. Certainly, the benefits of Reed's research on yellow fever fall on the positive side.

An often-cited negative example of participant consent is the Tuskegee Syphilis Study (Thomas & Quinn, 1991). The United States Public Health Service (USPHS) project followed 399 syphilis–infected, African American men for nearly 40 years. Although originally conceived as a 6-month study, the project developed a life of its own. Eventually, the USPHS actively prevented the infected men from receiving antibiotic therapy, misinformed them about the nature of the research, and prevented local physicians and other officials from informing them about their syphilitic condition. A *New York Times* article (Heller, 1972) finally brought the research to a halt. In response, the federal government's Tuskegee Health Benefit Program, which began in 1973, provided free medical and burial services to all remaining participants and their families. In 1997, President Clinton apologized to the seven surviving participants on behalf of the United States government. These two examples of the proper and improper use of informed consent set the stage for discussions of ethical principles in science and psychology.

We usually cover the Nuremberg Code that emerged from the Nuremberg trials that took place after World War II (Temme, 2003). The Nuremberg Code consists of 10 statements that outline ethical conduct for research with human participants. The Code was written after the "Doctor's Trial" of 16 defendants who had conducted research on humans in Nazi concentration camps. That research exposed unwilling participants to freezing temperatures, high altitudes, forced ingestion of sea water, and exposure to diseases and poisons. Many of the participants suffered pain and horrible deaths as a result. The research was clearly unethical. Even using the results of that research is considered unethical by the United States Environmental Protection Agency (Sun, 1988). Our students are usually unaware of the details of the Nazi experiments, and these accounts provide another teachable moment. The

Nazi atrocities show how terribly far humans can deviate from ethical behavior while ostensibly practicing science.

ETHICS IN PSYCHOLOGICAL RESEARCH

Psychology was fairly late incorporating and promulgating an ethical code governing its practitioners. Concern over the proper use of human participants in psychological research emerged strongly and persistently only after 1974 (McGaha & Korn, 1995). We concentrate on three examples, those of Milgram (1963), Zimbardo (1973), and Loftus (1997).

The research of Milgram (1963) and Zimbardo (1973) is well known. Loftus's (1999) characterization of her classroom project as research is less well known. For more information on Milgram and his research, see Blass (2000) and Harrison (2004). For more information on Zimbardo and his research, see Haney and Zimbardo (1998). For more information on Loftus, see the citation in the *American Psychologist* (2003) for the Distinguished Scientific Applications Award she received.

In class, we present the details of the three studies in chronological order. Briefly, Milgram sought to determine how much shock participants would agree to deliver to someone like themselves when instructed to do so by a researcher. Zimbardo's Stanford Prison Experiment created a situation in which initially similar participants changed radically within the span of a few days due to their roles as prisoner or guard, roles that had been assigned randomly. Loftus took a successful class demonstration and turned it into a compelling line of research. In the demonstration, one of her students successfully convinced his younger brother that he had been lost in a shopping mall when he was 5 years old, an event that never took place. In Loftus's subsequent research, she demonstrated that such false memories could be reliably created in a small percentage of participants. Later, however, she mischaracterized the classroom demonstration as a research project and ran afoul of critics (Crook & Dean, 1999). The reason we present these studies chronologically is that they illustrate how psychological research ethics evolved over time.

When Milgram began his research in the early 1960s, ethical concern for informed consent and knowledge about the power of the social role of the experimenter were far from understood. Indeed, the publication of Milgram's research promoted the development of ethics rules designed to protect participants and to more fully inform them prior to their participation (see Baumrind, 1964 and Milgram, 1964). By the time Zimbardo conducted the Stanford Prison Experiment in 1971, psychologists were more concerned and cognizant of the ethical issue of informed consent. To protect participants, many institutions had established institutional review boards (IRBs). The Stanford IRB reviewed Zimbardo's protocols and approved them after suggesting such improvements as keeping the fire extinguishers in the building.

The ethical issue raised by Zimbardo's research was whether to stop an ongoing and obviously successful project because of ethical concerns. Placing undergraduates into the roles of prisoner and guard worked all too well. As a researcher, Zimbardo wanted to continue the project, but as an observer of a microworld gone wrong, he knew that he had to put a stop to it. Later, he admitted that he had let the study continue too long (O'Toole, 1997).

Loftus was working in a more developed ethical environment than Milgram or Zimbardo. Loftus's "lost in the mall" research (1997) was inspired by a 1991 classroom exercise. Earlier, her attempts to have students convince friends or family that their memories for recent events were false had been unsuccessful. But several students reported success when she asked her students to convince someone that something that had never happened really did happen,. One student, Jim Coan, was spectacularly successful. With the cooperation of his parents, he had been able to convince his younger brother Chris (who was 14 years old at that time), that he had been lost in a local shopping mall when he was 5 years old, an event that never happened. After the classroom exercise, Coan and Loftus created a formal research proposal that the University of Washington IRB approved. However, at the 1992 meeting of the American Psychological Association, Loftus called Chris and others "pilot subjects" (Crook & Dean, 1999, p. 42) when, in reality, they were friends and family of her students who were participating in a class exercise. Loftus, too, had stumbled into a new and unexplored area of psychological research ethics: the relation between the classroom and the laboratory. Psychologists must distinguish between classroom teaching and research. The ethical concerns and responsibilities of these two arenas are different, as we note subsequently. These three historical examples prepare students for an examination of psychology's *Ethical Principles of Psychologists and Code of Conduct* (American Psychological Association, 2002).

PSYCHOLOGY'S ETHICAL PRINCIPLES

Our efforts to teach psychological research ethics always involve the American Psychological Association's (APA; 2002) publication, the *Ethical Principles of Psychologists and Code of Conduct* (referred to henceforth as the *Ethics Code*). In the 5th edition, teaching and research are separated into two sections. New ethical standards related to research in psychology are now in the sections on informed consent (8.02b, intervention research using experimental treatments), debriefing (8.08c, minimizing harm due to research procedures), and sharing research data for verification (8.14b, limits to the use of shared data for reanalysis). Differences between the 1992 (4th edition) and 2002 versions of the *Ethics Code* confirm the dynamic nature of research ethics in psychology.

Structure of the Ethics Code

Although we make no effort to teach the entire *Ethics Code* in our classes, we do emphasize its structure, which includes 5 broad general principles and 10 specific ethical standards. It covers nearly all of the ethical situations in which psychologists might find themselves. We begin by emphasizing Ethical Standard 4.01: Maintaining Confidentiality. Doing so, we believe, helps set a serious tone for future discussions of psychological research ethics. Maintaining confidentiality is also quite valuable outside of research settings. However, in class, our major concentration is on Ethical Standard 8: Research and Publication.

Our coverage of Ethical Standard 8 has three elements. One element is to provide and assign a copy of Ethical Standard 8 and follow this with an open-book test on a number of ethical scenarios. Students read the scenarios and decide whether they are ethical. If a scenario describes an unethical behavior, the student must cite the appropriate standard and propose a remedy. An example of one of our tests appears in Table 11.1.

The second element of our coverage of Ethical Standard 8 is to reconceptualize it as two separate, but related, components. The first component is *Responsibilities to Participants* and the second component is *Responsibilities to Psychology*. These two components help students better understand the ethics of psychological research. We will describe the two divisions in greater detail.

The third element of our coverage of Ethical Standard 8 is animal research. Only rarely do our students conduct animal research projects. However, we believe that all students should know and understand the ethics of research with animals.

Responsibilities to Participants

Under this heading, we place major emphasis on the following ethical standards:

Informed Consent (8.02)
Dispensing with Informed Consent (8.05)
Debriefing (8.08)
Deception in Research (8.07).

These four standards are important because they deal with the interactions between researchers and participants. Under informed consent, we reaffirm our earlier historical commitment to the standard by closely examining all eight subsections of the standard. We follow up by covering the three situations in which researchers may dispense with informed consent. The importance of debriefing participants is our next point of emphasis. Knowing why informed consent is necessary, when researchers can dispense with informed consent, and the importance of

TABLE 11.1
Ethics Scenarios and Answers

	Scenario	*Answer*
1.	A physiological research project using animals is underway when the researcher discovers that there is no anesthetic. The researcher proceeds with the study without anesthesia.	Unethical; 8.09 (f)
2.	A graduate student conducts a research project. Later the local IRB discovers that the graduate student did not submit a prospectus to the IRB prior to starting the study.	Unethical; 8.01
3.	A reviewer sees a manuscript and likes the ideas it contains. This stimulates the reviewer to begin a new personal research program along the same lines as the ideas in the manuscript. Later, when the reviewer publishes her new research, she cites the previous research she reviewed.	Ethical
4.	A study involving deception is planned but the outcome is well known and alternative non–deceptive methods have also been published in the literature.	Unethical; 8.07 (a)
5.	A study is conducted and later published in 2005. When a similar study is planned by the same researcher, a reanalysis of the 2005 data reveals serious errors. The researcher does not correct the errors.	Unethical; 8.10 (b)
6.	A department chair arranges for space for a research project and no more. When the study is published, the chair is listed as the fourth author.	Unethical; 8.12 (b)
7.	A study was conducted with a sample of high school dropouts. The informed consent form was adequate but 90% of the participants could not understand the language in the statement because it contained long sentences and unfamiliar words.	Unethical; 3.10 (a) or 8.02

debriefing are important elements of psychological research. Using or not using deception requires a decision that is based on a responsibility that researchers have to participants. Students should know that deception is permissible only when proposed research is likely to produce valuable results and when there are no nondeceptive methods available. We want our students to know that they should not use deception in research without justifying it.

We place lesser emphasis on the following ethical standards that relate to participants:

Informed Consent for Recording Voices and Images in Research (8.03)
Client/Patient, Student, and Subordinate Research Participants (8.04)
Offering Inducements for Research Participation (8.06).

Ethical Standards 8.03, 8.04, and 8.05 relate more to potential ethical problems created by technology, status, and magnitude of payment than they do to basic ethical concerns. Still, they fit under the umbrella of responsibilities to participants. When our students plan and conduct their research later in the course, we expect them to adhere to all of the ethical standards related to participants.

Responsibilities to Psychology

Under this heading, we place major emphasis on the following ethical standards:

Reporting Research Results (8.10)
Plagiarism (8.11)

We believe that emphasizing these two standards helps our students understand the seriousness of proper conduct of research. Presenting research results honestly and completely is a methodological and ethical requirement. Ethical Standard 8.10 is important because it clearly states the importance of honestly obtaining, reporting, and, if necessary, correcting results. Method and ethics come together in this standard.

We teach students about plagiarism and how to avoid it. Murray (2002) noted that many students are simply unaware that they are plagiarizing and that professors do not always provide a clear definition of *plagiarism*. We use Martin's (1994) analysis of plagiarism, which identifies five categories of plagiarism: word-for-word plagiarism, paraphrasing, secondary source plagiarism, idea plagiarism, and institutional plagiarism. Students, he found, are often guilty of infractions in all categories except institutional plagiarism. We take great care to define plagiarism and even greater care to prevent it.

Szuchman (2005) instructed students to avoid committing plagiarism when paraphrasing. Her principal advice is to avoid paraphrasing sentences and instead to paraphrase paragraphs. Doing so will make one less likely to plagiarize. She also covered other aspects of plagiarism such as using another author's examples or thematic outline. Both constitute plagiarism. Ultimately, she (and we) fall back on a simple rule: give credit. Giving credit means that students cite in their reports the sources that they use. We place lesser emphasis on the following ethical standards that relate to responsibilities to psychology:

Institutional Approval (8.01)
Publication Credit (8.12)
Duplicate Publication of Data (8.13)
Sharing Research Data for Verification (8.14)
Reviewers (8.15)

Most of these standards apply to the handling of research reports and their publication. The vast majority of our students do not submit their personal research for publication. However, Ethical Standard 8.01 notes that research is conducted most often in social settings that are governed by rules and regulations. Such regulations come directly from our respective schools' IRBs. We believe it is important for our students to know who serves on our IRBs as well as to know why our schools have rules governing the conduct of research in the first place.

Animal Research

As two psychologists trained in the comparative tradition, we believe that students should know what constitutes the ethical treatment of animals even if they never conduct animal research. Thus, we cover Section 8.09, Humane Care and Use of Animals in Research, in our courses. We use the APA's booklet, *Guidelines for Ethical Conduct in the Care and Use of Animals* (available at http://www.apa.org/science/anguide.html). Both Section 8.09 and the APA booklet discuss how to obtain animals, care for them, use them, and dispose of them. Animal research, too, is only ethically permissible when valuable results are likely and cannot be obtained without using animals. In sum, we end up covering all of Section 8 of the *Ethics Code*. However, we avoid teaching it from the top down or having students memorize it.

SOCRATIC CLASSROOM DEBATES

At the 2004 Symposium on Best Practices in Teaching Statistics and Research Methods in Psychology sponsored by Kennesaw State University, we demonstrated an ethical scenario that we use in class. The scenario, which involves secondhand smoke outside of campus buildings, appears in Table 11.2. At least three prominent ethical issues are built into the scenario in Table 11.2.

1. The health hazards of secondhand smoke,
2. Whether researchers obtained institutional permission to apply or to remove the "No Smoking" signs.
3. Whether the researchers obtained IRB approval (from *both* institutions) prior to conducting the research.

This scenario in Table 11.2 produces a lively debate in class, and it did so also at the Best Practices meeting. Because the deleterious effects of secondhand smoke are well known, audiences quickly seize upon the ethical issue of the researchers causing potential harm to participants at School A, where the existing signs prohibiting smoking near the doors were removed. (Later, we use Section 8.02a (2) to make the point that this situation requires informed consent because of the potential health

TABLE 11.2

Smoking in the Presence or Absence of "No Smoking" Signs

In this field experiment smoker compliance to newly re-moved or newly installed "No Smoking" signs will be observed. Buildings on two campuses will be used. On one campus, School A, smokers must not smoke within 25 feet of any campus building. Signs at School A inform smokers of this regulation. On the other campus, School B, smokers may smoke just outside of any campus building. There are no signs prohibiting smoking outside of the buildings at School B.

In the first part of the experiment, smokers will be observed on each campus to determine their average distance from the building. In the second part, signs will be removed or installed at each campus. At School A, the signs prohibiting smoking within 25 feet of any building will be temporarily removed. At School B, new signs prohibiting smoking within 25 feet of any building will be installed. Again, smokers will be observed to determine their average distance from the building. In the final part of the experiment, the signs on each campus will be returned to their original positions (Campus A) or removed (Campus B). The final data collection will measure smokers' average distance again. Because data will be collected by naturalistic observation, no informed consent by smokers is required.

risks.) Audiences, however, do not initially raise the question of whether the researchers have the right to add or remove signs until someone asks about the role of the IRBs. Being Socratic, we turn their questions around. We ask, "What if the IRBs had not been previously consulted?" "Unethical," is their reply. We then ask them to consider whether the IRBs should have approved the research in the first place. Usually, they reply, "No, the IRBs should not have approved the research because of the potential health risks." A minority, however, contend that the additional health risk of the secondhand smoke is essentially zero. Sometimes, we ask audiences to suppose that the IRBs had rejected the proposed research, but that the researchers had gone ahead and conducted it anyway. This tactic raises the question of sanctions for failure to obtain IRB approval.

An easy way to create scenarios for class use is to use news reports. Recent events such as the United States Supreme Court's decision to prohibit the use of marijuana as an analgesic drug or the federal government's ruling that allows the Department of Defense to administer drugs to military personnel without their consent (Files, 2004) can be turned into engaging scenarios. Note that creating successful scenarios requires situations that are not clearly unethical and that possess some resolvable ambiguity.

ETHICS IN STUDENT RESEARCH

Our research methods classes are structured around student research. Our students identify a research problem and then prepare a prospectus that consists of: Title Page, Introduction, Method, Expected Results, and References. After discussions with us, students present their prospectus to the class, which acts as an IRB committee. The class IRB receives copies of each prospectus before the presentations. At the class IRB meeting, student researchers present a short oral summary of their proposed research followed by ethical and methodological questions from the class IRB. Typically, the entire process takes about 30 min per prospectus. Afterward, the class IRB votes whether to allow the research to continue. Often, the class IRB recommends changes and holds another meeting after the changes have been made. The ethical and methodological lessons learned in class must later be demonstrated in the context of each student's research project. We find that nearly all of our students fully understand the requirements of the *Ethics Code* by the time they present their research to the class IRB.

Neither of our schools requires that students submit their research proposals to our institutional IRBs. Of course, some schools require that IRBs review all research, including student research. Kallgren and Tauber (1996) discussed how they handled their school's requirement that their IRB review student research. Their students reported that they viewed the IRB examination positively and believed that the process improved their research.

SUMMARY

We teach ethics throughout our research methods courses. We do not see ethics as just another topic in the course. Instead, we view ethics as an integral part of research that cannot be teased out or added onto later. We attempt to teach students that research is more than having an idea, investigating it, and then reporting it. By covering the general history of ethics and the history of ethics in science and psychology, we hope to dispel students' naive views about the relationship between ethics and research. One of the critical points we teach is that research ethics evolve. Our psychological examples are designed to show how landmark research not only revealed new data but also revealed new and unexpected ethical crises that needed resolution. Happily, we can report that nearly all of our students do learn why ethics are important to research, and they design research projects that are ethically sound. We expect our students to understand that "unethical research" is an oxymoron. Psychological research must be ethical.

AUTHOR'S NOTE

Please direct all correspondence to Edward P. Kardas, Department of Psychology, Southern Arkansas University, Magnolia, AR 71754–9199, e-mail epkardas@saumag.edu, or telephone (870) 235–4231.

REFERENCES

American Psychological Association. (2002). *Ethical principles of psychologists and code of conduct.* Washington, DC: Author.

Baumrind, D. (1964). Some thoughts on ethics of research: After reading Milgram's "Behavioral study of obedience." *American Psychologist, 19,* 421–423.

Blass, T. (2000). The Milgram paradigm after 35 years: Some things we now know about obedience to authority. In T. Blass (Ed.), *Current perspectives on the Milgram paradigm* (pp. 35–59). Mahwah, NJ: Lawrence Erlbaum Associates.

Crook, L. S., & Dean, M. C. (1999). "Lost in a shopping mall"—A breach of professional ethics. *Ethics and Behavior, 9,* 39–50.

Elizabeth F. Loftus: Award for Distinguished Scientific Applications of Psychology. (2003). *American Psychologist, 58,* 864–867.

Files, J. (2004, December 19). Defense Dept. asks to resume anthrax shots. *The New York Times,* p. 36, Sec. 1.

Haney, C., & Zimbardo, P. (1998). The past and future of U.S. prison policy: Twenty-five years after the Stanford Prison Experiment. *American Psychologist, 53,* 709–727.

Harrison, W. (2004). The man who shocked the world: The life and legacy of Stanley Milgram. *Personnel Psychology, 57,* 1081–1084.

Heller, J. (1972, July 26). Syphilis victims in U.S. study went untreated for 40 years; syphilis victims got no therapy. *New York Times,* p. 1.

Kallgren, C. A., & Tauber, R. T. (1996). Undergraduate research and the institutional review board: A mismatch or happy marriage? *Teaching of Psychology, 23,* 518–537.

Loftus, E. F. (1997). Creating childhood memories. *Applied Cognitive Psychology, 2,* S75–S86.

Loftus, E. F. (1999). Lost in the mall: Misrepresentations and misunderstandings. *Ethics and Behavior, 9,* 51–60.

Martin, B. (1994). Plagiarism: A misplaced emphasis. *Journal of Information Ethics, 3,* 36–47.

McGaha, A. C., & Korn, J. H. (1995). The emergence of interest in the ethics of psychological research with humans. *Ethics & Behavior, 5,* 147–159.

Milgram, S. (1963). Behavioral study of obedience. *Journal of Abnormal & Social Psychology, 67,* 371–378.

Milgram, S. (1964). Issues in the study of obedience: A reply to Baumrind. *American Psychologist, 19,* 848–852.

Murray, B. (2002). Keeping plagiarism at bay on the Internet. *Monitor on Psychology, 33,* Retrieved November 19, 2005 from http://www.apa.org/monitor/feb02/plagiarism.html

O'Toole, K. (1997, January 8). The Stanford prison experiment: Still powerful after all of these years. *Stanford News,* 30.

Pierce, J. R. (2003). "In the interest of humanity and the cause of science": The yellow fever volunteers. *Military Medicine, 168,* 857–863.

Sun, M. (1988). EPA bars use of Nazi data. *Science, 240,* 21.

Szuchman, L. T. (2005). *Writing with style: APA style writing made easy* (3rd ed.). Belmont, CA: Thompson Wadsworth.

Temme, L. A. (2003). Ethics in human experimentation: The two military physicians who helped develop the Nuremberg Code. *Aviation, Space, & Environmental Medicine, 74,* 1297–1300.

Thomas, S. B., & Quinn, S. C. (1991). The Tuskegee Syphilis Study, 1932 to 1972: Implications for HIV education and AIDS risk education programs in the Black community. *American Journal of Public Health, 81*, 1498–1505.

Zimbardo, P. G. (1973). On the ethics of intervention in human psychological research: With special reference to the Stanford prison experiment. *Cognition, 2*, 243–256.

Chapter **12**

Upper Division Research Methods: From Idea to Print in a Semester

Kenneth D. Keith
University of San Diego
Jody Meerdink
Nebraska Wesleyan University
Adriana Molitor
University of San Diego

Research methods constitute a key aspect of a strong undergraduate psychology curriculum (McGovern, Furumoto, Halpern, Kimble, & McKeachie, 1991). Purdy, Reinehr, and Swartz (1989) found that most of the directors of top graduate programs in psychology considered research experience an important criterion for admission to graduate study. Nevertheless, although undergraduate research seems to have increased in recent years (Kierniesky, 2005), Bailey (2002) suggested that many psychology students come to us with applied career interests and may not recognize the educational role played by methodological and analytic skills. It is thus essential that we teach research methods effectively (Ware & Brewer, 1999) and that we actively engage students in the learning process (McGovern et al., 1991). In this chapter, we describe our efforts to achieve active engagement in development of research skills at the upper division undergraduate level. We do this by presenting our approach to teaching two subject-specific research methods laboratory courses—in developmental psychology and cross-cultural psychology—and a senior-level thesis program.

COMMON CHALLENGES

We, and many other psychology teachers whom we know, encounter several predictable challenges in guiding students toward successful engagement with the research process. Froese, Gantz, and Henry (1998) identified four typical difficulties faced by students: (a) poor selection of research topics, (b) inadequate conceptual skills, (c) lack of integration of ideas from research articles, and (d) difficulty in evaluation of research articles. Chamberlain (1986) proposed several library activities, a series of critical reading questions, and a number of research design exercises as means to develop student research skills, and Poe (1990) described a series of writing assignments aimed at fostering critical analysis and integration of material from published research. Other psychology teachers (e.g., Addison, 1996; Carroll, 1986) have described specific means to development of individual and group student research proposals and projects, and Kallgren and Tauber (1996) discussed the values, both ethical and pedagogical, of undergraduate experience with institutional review boards.

Our focus in this chapter includes our approaches to (a) guiding students through identification and refinement of research ideas, (b) review and organization of relevant literature, (c) development of a specific research question or hypothesis, (d) design and conduct of a study capable of casting light on the question(s) posed, (e) selection of analytic techniques appropriate to the data generated, and (f) preparation of a discussion conceptually appropriate to the specific project. The overall task, we tell our students, is to select a topic and design a study that can be conducted, analyzed, and reported by (usually) one researcher, working alone and without significant material resources, in a single semester.

We work toward this common goal via three types of experience. One of us (Keith) conducts a cross-cultural psychology laboratory that asks students to identify and investigate their own topics; another (Meerdink) directs senior theses, again with students who identify and pursue their own ideas; and the third (Molitor) directs a developmental psychology laboratory in which students design projects using archived data. In each of our departments, participation in an upper division research experience is a requirement for the psychology major.

RESEARCH METHODS IN CROSS-CULTURAL PSYCHOLOGY

At the University of San Diego, psychology students choose at least one upper division research methods laboratory from among 10 in various topic areas (animal behavior, biopsychology, clinical, cognitive, cross-cultural, developmental, health psychology, human memory, learning and behavior, social). These labs share some common aims, including fostering library research skills, student development of an independent research topic, and preparation of an APA-style research report. Each laboratory experience carries three semester units of academic credit. The cross-cultural laboratory requires, of course, selection of a topic and methods appropriate to research problems of cross-cultural interest.

Selecting a Topic

Early in the lab semester, students participate in a review of basic research methods and statistics, and they select and complete critical reviews of published studies in areas of possible interest to them. They also read and criticize published undergraduate student work (cf. Ware, Badura, & Davis, 2002). These activities help to build student confidence in their skills and to move them toward selection of a personal research topic of interest. This is a difficult time for the students; many are indecisive, and those with ideas often find that their first efforts result in topics that are far too broad (i.e., the stuff of large books) or too narrow (trivial or tired). And the perfectionists find it impossible to commit to any idea at all if there is any chance that a better one might yet exist. Engagement in discussion of research as a complex human behavior not unlike any number of others, combined with fairly early deadlines for topic selection and identification of a number of sources, helps these students settle on research problems that are both meaningful and manageable.

Two notions are effective in moving students along in this process: *writing what we know* and *the funnel effect*. The former dictates that students become intimately familiar with their sources, whether print or electronic, making themselves reasonably expert in the particular problem they have chosen. Then, one can argue, they will be able to write with some authority from a perspective in which they have confidence. The funnel effect alludes to the fact that, if a topic is too broad, much of the information will simply "run over" and be lost and that, if the topic is too narrow, the process will not be efficient and the flow of information will be insufficient to the task. The metaphor suggests the importance of finding a middle ground of sufficient breadth to be meaningful without becoming overwhelming. Henderson (2000) described a useful procedure that can be adapted for use by students in developing a reader's guide for reviewing and communicating literature in a particular research area; and Pyrczak (2003) developed a student guide to evaluation of research published in academic journals.

The Research Question

It is not uncommon for students, having selected a research area, to experience difficulty in formulating a meaningful research question. Beginning researchers find it helpful to realize they need not necessarily raise questions so new or unique as to be totally unexpected or startling, but that science progresses through discovery of orderly processes that build upon the prior work of others (Sidman, 1960). Thus, students must see that systematic replications or extensions of prior work are not only acceptable, but often more likely to produce meaningful outcomes than are attempts to generate wholly original questions unrelated to previous research.

A second major issue for beginning researchers, one addressed in this lab, is the problem of framing a research question or hypothesis in such a way that it is testable. Does the question suggest the nature of the observation, the psychometric instrument, or the type of interview query that might provide clear answers? Students practice this skill in the lab, identifying the connection between research questions and measures in previously published work and practicing formulating questions for research topics of interest to members of the lab group.

Research Design

Cross-cultural researchers rarely conduct true experiments, owing to the fact that cultural characteristics cannot be randomly assigned, and cross-cultural studies present other special challenges that students should understand in designing their investigations. For example, noncultural demographic equivalence, linguistic equivalence, and cultural response sets are among the design issues that cross-cultural researchers must consider (Matsumoto & Juang, 2004). Student projects cannot always control threats to their design, but students should nevertheless be aware of these threats and be prepared to discuss their possible effects.

Student-designed projects in this lab need not manipulate variables, but they must be empirical; that is, students must implement a data collection procedure that addresses their particular research question. Perhaps the most common type of study designed by students in the lab would be cross-cultural comparisons examining attitudes, beliefs, communication, knowledge, or behaviors (in such areas as health, play, parenting, or education). Although these studies may not have sufficient depth, precision, or control to allow conclusions about the causes of cultural similarities and differences, they can serve such scientific purposes as illuminating the limitations of traditional psychological knowledge (Matsumoto & Juang, 2004). In addition to multigroup comparisons, students also sometimes design correlational studies, surveys probing psychological characteristics of cultural groups, naturalistic observations of behavior in cultural settings, and factorial investigations (e.g., culture × gender).

Institutional Review Board

Students in the cross-cultural lab must become familiar with the university Institutional Review Board (IRB) rules and procedures, and evaluate, with the instructor, the relation between their own projects and those rules. Most projects designed by students in the lab are eligible for expedited or exempt status; however, if this is the case, students nevertheless learn to determine that status by reference to applicable rules and regulations and to complete the required IRB forms for their studies. They prepare informed consent forms that meet IRB standards (or determine why their project may be exempt from informed consent requirements); and upon completion of projects, they submit their raw data and consent forms to the instructor for storage consistent with IRB rules.

These requirements serve not only to ensure compliance with rules and regulations, but also to increase student awareness of the legal, ethical, and logistical contingencies engendered by the IRB process. As Brinthaupt (2002) pointed out, an understanding of the relationship between researchers and the IRB, and its effects on the research process, is an important aspect of the learning of student researchers.

Student Products

Working in groups, students in the cross-cultural lab review and comment on drafts of their colleagues' work, and each student (or, occasionally, a pair or small group of students) prepares an APA-style paper reporting the research conducted during the semester. Although some experience difficulty completing the work required by this lab in one semester, virtually all students produce a final paper based on their own research project. The students present a summary of the finished product in the classroom setting, and a number go on to present their work at local or regional conferences or in departmental colloquia, and some submit their work for possible publication in undergraduate journals (usually the *Journal of Psychological Inquiry* or the *Psi Chi Journal*).

SENIOR THESIS RESEARCH

Since the 1960–1961 academic year, each senior psychology major at Nebraska Wesleyan University has completed a senior research project. Like the cross-cultural studies described previously, these projects must be empirical in nature, and they must report research on a topic of the student's choosing. Students in this program get a bit of a head start on the single-semester research project in that they complete a one-unit preparation (an upper division course titled Introduction to Senior Research) during the semester prior to actually conducting the study.

Introduction to Senior Research

Introduction to Senior Research involves small groups (5 to 7 students) in a process of "brainstorming" research ideas and developing research plans. Students generate their own ideas with a faculty advisor as consultant. Students develop proposals as they prepare a literature review, submitting 2 to 4 article summaries per week related to their topics.

This requirement helps to keep students on track and discourages the tendency to stop upon finding a few good articles. These students also experience, of course, the usual difficulties associated with selection of a topic, integration of diverse sources, and articulation of a research question.

The final product for the Introduction semester is a strong literature review, as well as a possible research hypothesis or research question and a supporting rationale. Students also propose a methodology and potential plans for data analysis. The papers include reference lists and additional materials the student may have developed (informed consent forms, questionnaires, surveys, or debriefing plans). The successful student has, at the end of Introduction to Senior Research, a sound plan for a project to implement during the subsequent semester. The faculty advisor evaluates this document and retains it (or passes it on, if a different advisor will be overseeing the senior research project).

Senior Research

During the first week of classes, all senior research students and advisors attend a general meeting to review the Senior Research syllabus. Each advisor supervises the work of 4 to 6 students and meets individually with each student to review (and potentially revise) the proposal developed in Introduction to Senior Research. The student continues to build the literature review and refine the planned introduction to the paper. In the first few weeks the student also develops a proposal for the Psychology Department Research Review Board (PDRRB), the local version of an IRB. The student should submit this proposal to the PDRRB during the fourth week of the semester.

At about the same time, students are refining their planned methodologies and dealing with any other project requirements (e.g., school district approval if the project involves minor students). Students spend much of the remainder of the semester in data collection activities, meeting occasionally with advisors for updates. They must complete data collection by the end of the 11th week of the semester, although advisors generally encourage an earlier completion date for this phase of the project. Students complete data analysis with guidance from advisors and submit a draft of the final paper by week 13. It is common for advisors to see three or four drafts of the paper before the student submits the final version.

Student Products

During week 15, all senior research students participate in a poster presentation session. Each student presents a research poster and a brief oral presentation for the benefit of an audience of senior thesis advisors, departmental faculty, senior research students, and other psychology students (especially those who will be engaged in senior research in the subsequent semester). Students submit two copies of the final paper, along with consent forms; one copy remains in the departmental archives and the advisor makes final comments and returns the second copy to the student. In addition to the departmental poster session, some students present their research at local or regional conferences, and some submit their work to undergraduate psychology journals.

ARCHIVED OBSERVATIONAL DATA IN A DEVELOPMENTAL PSYCHOLOGY LABORATORY

An interesting aspect of a laboratory experience in developmental psychology is the opportunity to examine data obtained from minors. Unfortunately, swift IRB approval is not likely, and the review process often entails at least one resubmission prior to approval. Thus, students may find it difficult, if not impossible, to complete a project in a single semester—an obstacle that can be overcome through use of data archives.

The Concept

One alternative to a student-initiated project needing IRB review is the use of previously archived raw data. Archived video records, in particular, allow the faculty advisor to accelerate the research process in some aspects (IRB approval, participant recruitment, administration of conditions/tasks) while providing students experience in key tasks (observing participants, making measurements, obtaining reliability, entering data sets, selecting statistics, analyzing data, interpreting and reporting results, and writing discussions). Parental consent for use of data in this way is, of course, essential. This section suggests a series of steps to move students from start to finish in submitting a complete research report that centers around archived video records.

Getting Started—The First 4 Weeks

Students begin by reading a seminal article that discusses the question(s) the archived video data can also address in some way. The article need not be recent, but it should help students understand what questions researchers are asking in this area, as well as the rationale behind the questions. Then, working with the instructor, students begin to translate the general topic into a specific question (or questions) that will guide the students' search of the remaining literature. Students often experience difficulty thinking in terms of specific independent variables/predictors and outcomes—making the guiding role of faculty key to a coherent, focused literature review. Cone and Foster (1993) offered a practical guide for students getting started with their research tasks.

These authors addressed such helpful topics as finding literature, decision rules for inclusion/exclusion, critiquing research, organizing literature, and developing hypotheses. In addition, Dunn (1999) and Smith and Davis (2004) provided organized approaches to the research process, with a particular focus on undergraduate students.

Students begin to provide summaries and evaluations of articles as soon as possible. A minimum of 10 to12 empirical studies can help to achieve adequate coverage; requiring a minimum number to meet a recent date requirement will ensure that students review the latest scholarship. Summaries and evaluations should be completed in batches, with deadlines along the way, to avoid the student tendency to wait for a final selection of articles. This procedure also provides opportunity for constructive feedback on each batch, enabling students to address potential problems before preparing a final draft of the introduction to their paper.

A final major step in getting started is student preparation of a list capturing the various limitations (methodological problems, important unanswered questions, insufficient answers) of the studies they have reviewed. Because not all problems can be addressed by archived data, the instructor must explain the nature of the archived video data (e.g., 5-min segments of 18-month-old toddlers' free play when alone and with their mothers, 4-month-old infants during three conditions of adult emotion). Students can then review their lists of limitations and begin thinking about how the archived data might contribute to addressing any of these limitations (i.e., how it may improve on past designs or measures, or what addition to knowledge archived data could provide).

Generating Quantitative Data

Weeks 5 to 10 of a 14-week semester must address students' movement from observing raw archived data to generating analyzable data (e.g., scores). Video records should be available in a location (the library reserve desk is good) that allows unrestricted out-of-class access in a private setting. A second key ingredient to students' autonomous navigation of video records is a participant log that identifies (a) participant numbers, (b) location on DVD/CD/tape of a participant's record, (c) which participants to score (code) for reliability purposes, and (d) which participant segments to score for data. A participant log can be stored and checked out with the videos at the library reserve desk.

The instructor must train students to observe/score participant behavior as a step toward generating quantitative data. The generation of data requires a list of behavioral definitions and scoring criteria, students' independent review of definitions, viewing preselected participant segments, and reviewing participants' scores. Following training, students independently score a sample of preselected participants previously coded by a criterion rater (e.g., professor) to calculate reliability. Students learn how to calculate an appropriate reliability measure (e.g., Cohen's kappa) and experience the challenges associated with achieving adequate reliability. Following reliability assessment, students independently score a number of participants and analyze data. If students have not scored a sufficient number of cases to make data analysis feasible, the professor can pool student subsets for students to analyze.

Analyzing Data and Wrapping Up

The lab includes a general review of statistics and selection of appropriate statistics, but students still need a good deal of practical help to navigate such resources as SPSS or other software for data analysis. Cronk (2003) provided many visuals (sample drop-down menus, sample dialogue boxes, sample output sections) and guided students through menu selections, output, and reports. As students analyze their data, the instructor provides any missing information about the nature of the archived data. Students receive this information orally to ensure that they practice skills in writing the method section of their reports. And, although there are constraints inherent in the nature of archived data and preselected behavioral measures, students remain free to produce thoughtfully their own hypotheses and rationale, as well as to discuss their sense of the implications of findings, limitations, and future research directions.

Cone and Foster (1993) provided practical guidance as students complete each major component of their report (introduction, method, results, discussion), and a checklist helps remind students of each step to complete along the way and can provide the basis for a grading rubric for use by the instructor. The result is a complete research paper, produced by students who might not otherwise have had access to young children, and embodying elements that would be missed if students only wrote proposals.

ASSESSMENT AND BENEFITS

The activities described here have potential to develop or enhance student skills in all the domains of scientific inquiry identified by Halonen et al. (2003): descriptive skills, conceptualization skills, problem-solving skills, ethical reasoning, scientific attitudes and values, communication skills, collaboration skills, and self-assessment. These skills, we believe, make such research involvement an appropriate part of the capstone activities of a comprehensive undergraduate psychology curriculum. Further, we can use samples of student research writing to assess achievement of important departmental aims. At the University of San Diego, for example, our departmental goals include such student outcomes as (a) evaluating, critiquing, and designing research studies; (b) implementing and interpreting statistics; (c) writing and reviewing papers in APA style; and (d) learning how research methods are applied to the study of behavior. The written products of student research can provide appropriate benchmarks for assessing attainment of these goals. Finally, students in the programs we have described have been successful in sharing their work with a wider audience (e.g., Bausch, Girón, & Sepp, 2003; Downey, 2004; Kiel, 1997; Nguyen & Huynh, 2003; Verbeck, 1996; Wells, 2001; Witte, 200l), and evaluation of student writing has become a part of our departmental assessment plans.

Students sometimes come to the research experience with trepidation or with the fear that research is simply another obstacle in the path to graduation. We aim, in our laboratories, to lead students to an appreciation of the process. Skinner (1970) claimed that "I have conducted experiments as I have played the piano. I have written scientific papers and books as I have written stories and poems" (p. 17). We do not presume to have achieved this level of aesthetic appreciation of research in our students. We do hope, however, that our endeavors have helped our students to see that scientific investigation is a satisfying, engaging process that is accessible, educational, and enjoyable. We also hope that our experiences will be helpful to others in their efforts to introduce students to the pleasures of conducting research.

REFERENCES

Addison, W. E. (1996). Student research proposals in the experimental psychology course. *Teaching of Psychology, 23,* 237–238.

Bailey, S. A. (2002). Teaching statistics and research methods. In S. F. Davis & W. Buskist (Eds.), *The teaching of psychology: Essays in honor of Wilbert J. McKeachie and Charles L. Brewer* (pp. 369–377). Mahwah, NJ: Lawrence Erlbaum Associates.

Bausch, M., Girón, M., & Sepp, A. (2003). Individualistic and collectivistic attitudes toward marriage. *Journal of Psychological Inquiry, 8,* 11–15.

Brinthaupt, T. M. (2002). Teaching research ethics: Illustrating the nature of the researcher IRB relationship. *Teaching of Psychology, 29,* 243–245.

Carroll, D. (1986). Use of the jigsaw technique in laboratory and discussion classes. *Teaching of Psychology, 13,* 208–210.

Chamberlain, K. (1986). Teaching the practical research course. *Teaching of Psychology, 13,* 204–208.

Cone, J. D., & Foster, S. L. (1993). *Dissertations and theses from start to finish: Psychology and related fields.* Washington, DC: American Psychological Association.

Cronk, B. C. (2003). *How to use SPSS: A step-by-step guide to analysis and interpretation* (3rd ed.). Los Angeles: Pyrczak Publishing.

Downey, K. B. (2004). At the playground: Cultural differences between Mexican and Euro-American children's play behavior. *Journal of Psychological Inquiry, 9,* 7–13.

Dunn, D. S. (1999). *The practical researcher: A student guide to conducting psychological research.* New York: McGraw-Hill.

Froese, A. D., Gantz, B. S., & Henry, A. L. (1998). Teaching students to write literature reviews: A meta-analytic model. *Teaching of Psychology, 25,* 102–105.

Halonen, J. S., Bosack, T., Clay, S., McCarthy, M., Dunn, D. S., Hill, IV, G. W., McEntarffer, R., Mehrotra, C., Nesmith, R., Weaver, K. A., & Whitlock, K. (2003). A rubric for learning, teaching, and assessing scientific inquiry in psychology. *Teaching of Psychology, 30,* 196–208.

Henderson, B. B. (2000). The reader's guide as an integrative writing experience. *Teaching of Psychology, 27,* 130–132.

Kallgren, C. A., & Tauber, R. T. (1996). Undergraduate research and the institutional review board: A mismatch or happy marriage? *Teaching of Psychology, 23,* 20–25.

Kiel, K. J. (1997). Effects of self-efficacy sources on pain tolerance in a cold pressor test. *Journal of Psychological Inquiry, 2,* 13–17.

Kierniesky, N. C. (2005), Undergraduate research in small psychology departments: Two decades later. *Teaching of Psychology, 32,* 84–90.

Matsumoto, D., & Juang, L. (2004). *Culture and psychology* (3rd ed.). Belmont, CA: Wadsworth/Thomson Learning.

McGovern, T. V., Furumoto, L., Halpern, D. F., Kimble, G. A., & McKeachie, W. J. (1991). Liberal education, study in depth, and the arts and sciences major—Psychology. *American Psychologist, 46,* 598–605.

Nguyen, A. M., & Huynh, Q. L. (2003). Vietnamese refugees and their U.S.-born counterparts: Biculturalism, self-determination, and perceived discrimination. *Psi Chi Journal, 8,* 47–54.

Poe, R. E. (1990). A strategy for improving literature reviews in psychology courses. *Teaching of Psychology, 17,* 54–55.

Purdy, J. E., Reinehr, R. C., & Swartz, J. D. (1989). Graduate admissions criteria of leading psychology departments. *American Psychologist, 44,* 960–961.

Pyrczak, F. (2003). *Evaluating research in academic journals: A practical guide to realistic evaluation* (2nd ed.). Los Angeles: Pyrczak Publishing.

Sidman, M. (1960). *Tactics of scientific research: Evaluating experimental data in psychology.* New York: Basic Books.

Skinner, B. F. (1970). An autobiography. In P. B. Dews (Ed.), *Festschrift for B. F. Skinner* (pp. 1–21). New York: Appleton-Century-Crofts.

Smith, R. A., & Davis, S. F. (2004). *The psychologist as detective: An introduction to conducting research in psychology* (3rd ed.). Upper Saddle River, NJ: Prentice-Hall.

Verbeck, A. (1996). Perceived attractiveness of men who cry. *Journal of Psychological Inquiry, 1,* 5–10.

Ware, M. E., Badura, A. S., & Davis, S. F. (2002). Using student scholarship to develop student research and writing skills. *Teaching of Psychology, 29,* 151–154.

Ware, M. E., & Brewer, C. L. (Eds.). (1999). *Handbook for teaching statistics and research methods* (2nd ed.). Mahwah, NJ: Lawrence Erlbaum Associates.

Wells, R. (2001). Susceptibility to illness: The impact of cardiovascular reactivity and the reducer–augmenter construct. *Journal of Psychological Inquiry, 6,* 12–14.

Witte, A. L. (2001). Cross-cultural study of pace of life on two campuses. *Journal of Psychological Inquiry, 6,* 84–88.

Part V

Integrative Approaches to Teaching Methods and Statistics

Chapter 13

Benefits and Detriments of Integrating Statistics and Research Methods

Andrew N. Christopher and Mark I. Walter
Albion College
Robert S. Horton
Wabash College
Pam Marek
Kennesaw State University

WHY COMBINE STATISTICS AND RESEARCH METHODS IN A SINGLE COURSE?

Context often provides a valuable cue for remembering events (Fisher, Geiselman, & Amadox, 1989) and verbal material (Bransford & Johnson, 1972), particularly if the material has not yet been mastered. The context in which psychology majors encounter statistics is when they are reading, planning, or conducting research, all of which are involved, to varying degrees, in research methods courses. Thus, a primary reason for integrating statistical training directly into the research methods course is to provide students with a relevant context to reinforce their memory of the statistics that instructors introduce. Rather than being potentially overwhelmed with learning one statistical technique after another (with only a few practice examples), students become acquainted with statistical tests as they gain knowledge of the continuum of research methods: observation, description, prediction, and explanation. Learning statistical techniques in this manner is likely to boost students' understanding of which statistics are best suited to specific research designs and what information can be garnered from different types of designs.

Related to the importance of context, a second reason for integrating statistics and research methods in a single course is the omnipresent difficulty students have in transferring problem-solving procedures from one context to another (Gick & Holyoak, 1980). If students take statistics as a separate course housed outside the psychology department, few examples, if any, from their statistics courses are likely to deal specifically with psychological research. As novices, these beginning students are likely to base future judgments of when to use specific statistics on surface similarity, or the "story line" of the problems in which they first encountered these statistics. Nisbett, Krantz, Jepson, and Kunda (1983) emphasized that statistical training can ameliorate people's tendency to inappropriately apply heuristics in reasoning. Yet, only more adept students readily recognize structural similarities and commonalties in the underlying procedures needed to solve problems (Novick, 1988). Thus, when students enter their methods courses, only those with the strongest skills are likely to recognize when it is appropriate to use a statistic that they learned previously. Given that most students would require considerable review, it appears more efficient to teach the statistics initially in conjunction with the research methods with which they are used. To facilitate instructors' ability to adopt this integrated approach, several textbooks (e.g., Davis & Smith, 2005; Furlong, Lovelace, & Lovelace, 2000; Heiman, 2001; Jackson, 2006) explicitly combine research methods and statistics, although textbook organization varies considerably.

A third reason for integrating statistics and research methods is to mirror the realities of reading, proposing, or conducting research outside the classroom. As methods instructors often stress, investigators typically formulate the analysis plan for a research study when they initially design the project. Perhaps even more important, the design of the research determines the type of statistics that investigators use. After researchers complete data collection, they cannot change a scale from dichotomous to continuous, nor can they alter the number of items on a scale. Yet, these types of methodological factors influence the selection of statistics. In quantitative research, statistics are paramount for determining whether a construct is reliable and whether the data answer the research question. Clearly, methodological considerations are inherently intertwined with statistical considerations, a compelling reason for teaching them both in the same course. In doing so, instructors also highlight the scientific nature of the methodological scaffold for psychological research.

We offer a three-pronged rationale for integrating research methods and statistics, based on the importance of relevant context, the difficulty of transfer, and the actual use of methodological and statistical skills outside the classroom. To foster scientific thinking, we contend that integration provides students with an opportunity to process information at a deeper level, to directly connect psychological theories to empirical tests of these theories, and to make longer-lasting connections between methodology and statistics. In the remainder of this chapter, we examine different techniques for accomplishing such integration.

Integrated research methodology and statistics courses are theoretically and pedagogically sound ventures, yet putting such a course into practice reveals a number of unique challenges. In the next two sections, we identify qualities that we regard as essential for the success of an integrated course and obstacles to that success. The lists of qualities and obstacles identified here include those things we regard as particularly noteworthy.

THE ESSENTIAL CONSIDERATIONS FOR INTEGRATION

Although there is no secret or magical combination of elements that will ensure a successful integrated research methods and statistics course, we discuss some considerations here that we believe can provide an effective structure on which to build. It is important to note that we do not try to provide even a small representation of those things that we would recommend for an integrated methods and statistics course (such a list would be quite lengthy and would be influenced by a variety of institution-specific factors, such as student skill level and faculty teaching loads). We confine our comments to what we regard as near-essential elements of the integrated course experience.

Instructional Time

The first, and possibly most obvious, consideration that seems essential for the success of an integrated course is adequate instructional time. We have found that teaching an effective integrated course demands more than double the amount of time that one takes for an isolated methods or statistics course. This fact may seem odd, because an integrated course seems to merely merge together two single courses. However, the course functions most effectively when one teaches the course over two semesters (or with double allocation during one semester) with a separate laboratory component (roughly 2 hr) each week. This additional instructional time is necessary because the integrated course asks students to go a step (or more) beyond what they would do in an isolated methods or statistics course; they must understand the information within each domain and then *synthesize* it across domains. For example, in a statistics course, a student must understand how to conduct an independent samples *t* test by hand. A research methodology course would instruct a student in between-subjects two-group research designs and in the benefits and detriments of both experimental and nonexperimental methods alike. In the integrated course students must master those same concepts and skills and be able to link the design to the analysis. The integrated course does not "cut corners" on the content of either statistics or research methods, but rather takes students to the next plane of understanding of how these two areas are intimately connected. That additional step of synthesis demands additional time. As mentioned, however, this extra time investment provides a context in which knowledge of one domain reinforces knowledge in the other domain.

Intergrating Methods and Statistics Instruction

Once appropriate instructional time is allocated, the question becomes how to use that time most effectively. A major challenge for an integrated methods and statistics course is how to move the course effectively from methodological to statistical instruction and back again. Such effective movement is the second component critical to the success of the course.

The task of integrating the material is not an easy or obvious one. There are a variety of schedules that may adequately accomplish the goal; however, the ideal scenario seems to be one in which discussion of a particular method or design and appropriate statistical procedures, both descriptive and inferential, are presented in succession, as a unit. This ideal scenario is complicated somewhat by the fact that most statistics texts present descriptive statistics first and then turn to inferential procedures. Indeed, there are few texts available that attempt to integrate research methods and statistics in a manner that even approximates the ideal scenario previously mentioned (a notable exception is Jackson's [2006] text). Having such a text to help guide the course would facilitate greatly the effective integration of material.

However, if one chooses traditionally formatted statistics and methods texts, the task of integration is not doomed. To approximate the ideal scenario previously described, one would simply need to front-load instruction in the basic properties of inferential statistics (e.g., *probability, sampling error, sampling distributions*). Placing these concepts at the beginning of the term would give students the basic understanding needed as they begin to apply inferential statistics, along with their descriptive compatriots, to particular research methods.

Presenting methods and appropriate statistics as a unit would be ideal for smoothing the sometimes rough transition back and forth between the two areas; however, we should acknowledge that the reality of the methods–statistics transition may take many different forms, each of which will have its pinnacles and pitfalls. As one example, the Wabash College course sequence has regularly included a first semester devoted only to research methodology and descriptive statistics (see Table 13.1), with the second semester of the course devoted to inferential statistics (see Table 13.2). Given this organization, students' first semester research experiences involve only descriptive statistical analysis and focus more heavily on the qualities of effective methodology. In our experience, such an organization is adequate, as it gives students a primary focus on methodology but also forces them to use simple statistics (e.g., means, medians, standard deviations, correlation coefficients) to address a research question. However, the organization also isolates the second semester course as, primarily, an inferential statistics course and does not achieve the seamless flow from methods to stats that appears ideal. As another example, at Albion College, the course sequence is structured such that during the first semester, students learn about nonexperimental methodologies and corresponding statistical analyses (see Table 13.3). In the second semester, students learn about experimental methodologies and corresponding statistical analyses (see Table 13.4).

TABLE 13.1

Class Schedule at Wabash College – First Semestera

Approximate Class Dates	*Topic*
Aug. 26	Introduction to the class
Aug. 31	How do you know things? Introduction of Mini-Project #1[b]
Sept. 2	Discussion of Mini-Project #1
Sept. 7	Types of Research & Mini-Project work
Sept. 9	Mini-Project #1 due, Intro to JMP
Sept. 14	Basic Statistics; JMP
Sept. 16	Playing with JMP
Sept. 21	Mini-Project #2 introduced[b]
Sept. 23	Writing Introduction Sections Validity and Measurement
Sept. 28	V & M catch-up/cont.
Sept. 30	Mini-Project #2 Work
Oct. 5	Mini-Project #2 Due. Discussion
Oct. 7	JMP practice
	Article review assigned
Oct. 12	Threats to Validity
Oct. 14	No class: Fall Break
Oct. 19	Article review due, Discussion
Oct. 21	Normal Curve- Standard Scores
	JMP
Oct. 26	Mini-Project #3 Introduced[b]
Oct. 28	Non-experimental designs
Nov. 2	Correlation and Regression; JMP
Nov. 4	Correlation and Reg., cont.
	Mini-Project #3 work
Nov. 9	Mini-Project #3 work
Nov. 11	Article review assigned
	Mini-Project #3 Due
	Discussion
Nov. 16	Experiments and Quasi-experiments
Nov. 18	Article review due, Discussion
Nov. 30	Intro to Inferential Statistics
Dec. 2	Catch-up and Review
Dec. 7	Presentations of Proposals[a]
Dec. 9	

[a]See CD for complete syllabi.
[b]See CD for descriptions of the mini-projects.

TABLE 13.2

Class Schedule at Wabash College – Second Semester[a]

Principle question/topic	*Class Dates*	*Specific Topic*
Organization, Review, and some Fundamental principles for inferential statistics	Jan. 15	Format, scheduling
	Lab Jan. 19, 20	No lab meeting; research meetings
	Jan. 20, 22	Review inferential statistics/z test
	Lab Jan. 26, 27	Review for JMP
Inferential statistics (I.S.) when you have a single group and one continuous outcome variable	Jan. 27	Confidence intervals
	Jan 29	No class meeting; assignment for one-sample t test
	Lab Feb. 2, 3	
	Feb. 3	One-sample T-test
I.S. for 1 IV, two-group research with a continuous outcome variable	Feb. 5	Independent samples t test
	Lab Feb. 9, 10	
	Feb. 10	Methods Assignment
	Feb. 12	Dependent-samples t test
	Lab Feb. 16, 17	
	Feb. 19	
	Lab Feb. 23, 24	First Exam!!!!!! First Lab Exam!!!
I.S. for multiple group research (1 IV); continuous outcome.	Feb. 24, 26	Analysis of Variance (ANOVA)
	Lab Mar. 1, 2	
I.S. when you have multiple IVs; continuous outcome.	Mar. 2, 4	Two-Way ANOVA
	Lab Mar. 15, 16	
	Mar. 16, 18	2-Way ANOVA; simple effects
	Lab Mar. 22, 23	
	Mar. 23	Review
	Mar. 25	Second exam!!!!
	Lab Mar. 29, 30	Second lab exam!!!!
		Work on projects! MAUPRC conference
	Mar. 30, April 1	Power
	Sat. April 3	Multiple regression
	April 6	
	April 8	Analysis of covariance (ANCOVA)
	Lab April 12, 13	
	April 13, 15	
	Lab Apr. 19, 20	
What if your outcome variable is not continuous?	April 20, 22	Chi-square
	Lab Apr. 26, 27	
	April 27, 29	Posters and wrap-up

[a]See CD for complete syllabi.

TABLE 13.3

Class Schedule at Albion College – First Semester[a]

Approximate Class Dates	*Topic*
Aug. 23–30	Course overview
	Introduction to statistics
	Introduction to methodology
Aug. 30	Ethics in research
Sept. 1	Physical trace and archival research
Sept. 6	Observational research
Sept. 8–10	Frequency distributions
Sept 13–15	Central tendency
Sept. 17–20	Variability
Sept. 22–27	Introduction to hypothesis testing
Minor Project #1: (the first group observational project with just a title page, abstract, methods and results) is due by 4:00 on Monday, September 27th.	
Sept 29 and Oct. 1	Exam #1 will cover all the material through this point
Oct. 4	Chi-square
Oct. 6	*z* scores
Oct. 13	Probability and the normal curve
Oct. 18	Survey Methods
Major Project #1 (the group observational study with all the parts of an APA paper included) is due by 4:00 on Wednesday, October 20th. Presentations of the group observational study will be in lab on the 21st or 22nd (depending on the lab in which you are enrolled). Approximate Class Dates	
Oct. 20–27	Correlation and regression
Nov. 1	Single-Subject Designs
Nov. 3–4	Exam #2 will cover all the material through this point
Nov. 8–10	Introduction to experimentation
Nov. 15	*t* tests with one sample
Minor Project #2 (the introduction to what will become major project #2) is due by 4:00 on Wednesday, November 17th.	
Nov. 17–22	Hypothesis testing with 2 independent samples
Nov. 29	Review/recap
Dec. 1–2	Exam #3 will cover all the material through this point.
Major project #2, (the individual survey project with all the parts of an APA paper included) is due by noon on December 13th. The time of the poster presentations of the survey project will be announced in class.	

[a]See CD for complete syllabi.

TABLE 13.4

Class Schedule at Albion College – Second Semester[a]

Approximate Class Dates	*Topic*
Jan. 10–Feb. 3	Review of RDA 1
Feb. 3–4	Exam #1
Feb. 8–10	Repeated measures designs
	Repeated measures *t* tests
Feb. 15–Feb. 22	Between subjects designs
	One-way ANOVAs
Feb. 24	Research Project #1 Paper Due
Feb. 24–Mar. 1	One-way repeated measures ANOVA
Mar. 3–4	Exam #2
Mar. 15–Mar. 22	Factorial designs
Mar. 24–Mar. 29	Quasiexperimental designs
Mar. 31–Apr. 5	Multivariate designs
Apr. 7–8	Exam #3
Apr. 7–Apr. 28	Presentation of individual projects
Apr. 28	Research Project #2 Paper Due

[a]See CD for complete syllabi.

In short, the manner of integrating methodology and statistical material will be affected by a variety of factors, including, but not limited to, the text(s) chosen, the expectations for the course within the departmental curriculum, the nature of the students, and the individual style of the instructor. We present "ideals" for consideration and as bases off of which to adjust given particular needs and resources.

Using Statistical Software, Deemphasizing Hand Calculation

As noted previously, the integrated methods and statistics course makes extensive demands on students' and instructors' time and academic resources. There is an exceptional amount of material to cover, synthesize, and use; thus, there are places in which instructors must make choices about which material to cover and how to cover it. One such choice that makes good theoretical and pedagogical sense is to deemphasize hand calculation of advanced statistical procedures, and instead, emphasize mastery of statistical software (e.g., SPSS, SAS, JMP) for those purposes.

Few psychologists that we know perform even a simple *t* test, much less a two-way ANOVA, by hand. Psychologists almost certainly use a statistical program to enter and analyze data, focusing energy more on the questions of what statistical procedure is appropriate, how to perform the procedure with the software, how to interpret the output, and what follow-up tests are needed. As instructors train students in psychological science and work to give them a realistic experience of what psychologists do and how science proceeds, we contend that instructors should spend their scarce instructional time on those skills that psychologists actually use. Of course, it is also important to teach students the fundamental operations of statistics (e.g., sampling distributions, standard error, partitioning variance). In fact, we are committed to the notion that students should understand how statistics operate and, in turn, what those statistics really tell us about the data. We suggest only that undergraduate psychology students may be better served by, for instance, learning how to produce an ANOVA summary table with software and then interpret it, rather than repeatedly subtracting raw scores from the group mean, squaring them, adding them all up, and the adding all of those sums across conditions (as one does when computing the within-participants sum of squares by hand).

To be optimally effective, instructors must work to balance effectively instruction in (a) the basic operations of statistics and (b) their practical use and interpretation. Based on our own experiences, we recommend that students perform all descriptive statistics and z and t tests by hand and also learn how to perform the tests with software. Beginning with one-way ANOVA, and continuing with the more advanced procedures, we recommend that instructors spend time to explain the underlying logic of the calculation (e.g., what are the between and within-groups variances and how do they combine to form an F-statistic?) and then move quickly to working with software for calculation and interpretation. In our view, emphasizing the use of statistical software, especially for advanced statistical procedures, provides a more realistic depiction of the tasks that confront scientific psychologists and the skills that such psychologists must have.

Certainly, emphasizing instruction in statistical software places an additional demand on the course: access to computers. The ideal situation is for each student to have access to a computer and adequate desk space to allow individual written work. Such access optimizes the flexibility of the course; any day of the semester may be devoted to either statistical or methodological work without any inconvenience. This flexibility is ideal, but if not possible, instructors may plan their syllabi to include practice with statistical software during computer laboratory sessions and use a noncomputerized classroom for other meetings.

The Student Research Project: From Beginning to End

As a final critical component for an integrated methods and statistics course, we turn to a specific course requirement that we hold in the highest esteem: the student research project. Psychological research involves more than just methodology, more than just statistics. After psychologists make an observation, develop a theory, devise a method to test the theory, state a hypothesis, and collect data (all topics covered in most research methodology courses), the psychologist must then analyze the data to answer some question (the issue dealt with in statistics courses). In the absence of any one of those steps, the research process is incomplete and artificial. We recommend that students experience the full research process in the integrated course.

There are many ways to encourage the experience of this process. At Wabash College and Albion College, in fact, we have different means for students to complete this vital process (see this volume's CD for information from both schools). Despite these differences, we recommend including student participation in at least one research project from its origin to its completion. Students should work to develop a theory, state hypotheses, devise a method, collect and analyze data, and report results on a single research question. In our classes, such experiences have often involved student-generated research ideas, but instructors can also assign research topics to students, as long as those topics are selected carefully for potential student interest and ethical compliance.

Regardless of how the research ideas arise, students commonly regard these complete research experiences as the most valuable instructional tool in our courses, and we regard them equally favorably. The projects force students to "get their hands dirty," so to speak, and give them responsibility for and ownership of their learning. A project allows students to learn from mistakes and the inevitable difficulties that arise. In fact, it is these difficulties, and the null findings that often result from these projects, that teach students critical lessons: (a) research is a process, a process that often demands assessment and revision; and (b) asking and testing good questions, rather than getting good results, are the truly critical goals of research.

Of course, it is possible to teach an integrated methods and statistics course without including a student research project; however, we find the complete student research project to be nearly unparalleled in its elegant value for the students. It focuses student engagement onto a single research question, providing natural cohesiveness and hierarchical organization to the student's education of many different ideas. When completing such a project, students are in the process of science and must use methodological and statistical tools for the purpose intended, not to receive a grade, but to answer a scientific question.

Concluding Remarks

The components listed here are not exhaustive of the considerations when embarking on an integrated methods and statistics course, yet we consider them critically important to the success of such courses. We believe that integrated courses can facilitate student learning in both research methods and statistics (relative to isolated courses), especially if instructors heed these recommendations. Each of these recommendations works for the good of the student and, albeit indirectly, for the good of the instructor charged with their guidance. We turn now to a discussion of the unique obstacles that integrated research methods and statistics courses present.

POTENTIAL OBSTACLES WHEN INTEGRATING STATISTICS AND RESEARCH METHODS

Although there are numerous pedagogical advantages to teaching statistics and methods in a combined sequence, several obstacles also exist. Although we believe that the aforementioned benefits outweigh the obstacles, it is important to keep a few obstacles in mind as a department decides the direction it wants to take when teaching statistics and methodology. Consideration of these obstacles prior to starting an integrated statistics and research methods course may help a department prevent problems to student learning. The obstacles can be categorized as those that are faced by a course instructor or a department.

Instructor Considerations

From an instructor's perspective, one of the first hurdles that needs to be addressed is the expectation that teaching the combined course will be more time consuming and difficult than teaching two separate courses. Indeed, it is not uncommon for faculty who teach statistics or research methods separately to be reluctant to integrate them into a combined sequence for just these reasons. We argue that given proper planning and departmental support, neither of these considerations represents an insurmountable obstacle. Regarding the first concern, it is true that teaching the combined course can, at times, be more of a time commitment than teaching each course separately. We have found that the greatest time commitment comes in the form of helping students analyze their data and in grading and commenting on full-scale projects and resultant APA-style papers that are typically assigned in this course. The time commitment is such that the enrollment in the combined course needs to be quite small. In our experience, 25 students is a reasonable class size, with two laboratory sections of 12 to 13 students in each. If a department follows the format of having an instructor teach one lecture section with two smaller lab sections, it should count as teaching two classes toward the instructor's teaching load. This format has the disadvantage of taking faculty away from other classes that they could be teaching, but we strongly believe the increased workload makes this necessary.

Regarding the issue of whether the combined sequence is harder to teach, it is true that when teaching statistics and methodology separately, the workload and the grading are spread out over two topics. When they are combined, the instructor may be faced with separate assignments assessing the student's knowledge of statistics and methodology as well as combined assignments (e.g., a full APA-style empirical paper) integrating statistics and methodology. In our experience, though, most instructors naturally find themselves integrating the two courses even when they are supposed to be separate. It is often difficult to teach statistics without reference to the research designs for which they are appropriate. To a lesser degree, teaching methodology begs the question, "What do I do with the data once I've collected it?" We would argue that although there may be more work for the instructor, this work does not have to be excessive and, in our view, has the advantage of being more pedagogically sound. That said, it is necessary to include a caution regarding the workload. We suggest, from personal experience, that it is possible to give students too many research projects in the course. With students learning numerous research methods and the statistics to analyze them, it is tempting to want to have students demonstrate their knowledge of these topics by having them collect and analyze data as well as write their study in a full (even if brief) APA-style paper. We find that having students conduct one, or perhaps two, studies per semester provides an optimal learning experience while not overwhelming the instructor with work. We have also assigned "miniprojects" that allow students to gain methodological and statistics experiences prior to conducting their own empirical studies (see the CD for a description of these tools). In sum, we suggest that the numerous pedagogical advantages outweigh the increased time and workload of teaching the combined sequence.

Departmental Considerations

When considering the introduction of an integrated research methods–statistics course, the department confronts questions related to laboratory space, transfer credits, and concerns of other departments. First, when integrating the two courses, we believe it is necessary to include a laboratory component as part of the course. Given that resources are not unlimited, one hurdle that needs to be addressed is that of scheduling laboratory space. As the number of sections of the course increases, it becomes increasingly difficult to schedule the laboratory component. Using laboratory time is more of an issue when there are other courses that are placing a demand on the available laboratory space.

Another departmental issue that can arise when teaching this course concerns transfer credits. It is possible that students will want to have the integrated course waived because they have taken statistics and/or research methods (or perhaps an integrated course) at another institution. It is up to the individual department to decide if students have to take the combined sequence if they have only had statistics but not methods or vice versa. We argue that the combined course should be required if the student has not had both a statistics and a methods course, even though some of the material will overlap with the transfer credit. Although we see the combined course as advantageous to taking each separately, we do not suggest that students be required to take this course if they have previously had each course separately.

When a psychology department decides it will teach statistics and methodology in a combined course, the repercussions can extend beyond the department. It is not uncommon for schools (particularly small schools) to require psychology majors to take statistics through the mathematics department. When the mathematics department no longer has psychology students enrolling in its statistics course, it could have a substantial negative impact on their enrollment numbers. It is important to consider this issue when planning on whether to integrate the two courses. We suggest that pedagogy should be the primary concern when making this decision.

CONCLUSION

There are sound pedagogical and pragmatic reasons to teach statistics and research methods in a single course. In fact, we have found that teaching an integrated course is more natural than trying to separate the two topics in different classes. On the accompanying CD, we have included several resources, including syllabi, descriptions of class projects, and SPSS datasets with suggestions for use that can be used to help develop an integrated statistics and research methods course. Indeed, as is the case with most every other aspect of teaching, there is no "one size that fits all" when trying to integrate statistics and research methods. We are reminded of Halonen's Corollary, "What works well for some teachers fails miserably for others" (Halonen, 2001). Such is the case when integrating statistics and research methods.

REFERENCES

Bransford, J. D., & Johnson, M. K. (1972). Contextual prerequisites for understanding: Some investigations of comprehension and recall. *Journal of Verbal Learning and Behavior, 11*, 717–726.

Davis, S. F., & Smith, R. A. (2005). *An introduction to statistics and research methods: Becoming a psychological detective.* Upper Saddle River, NJ: Pearson Prentice-Hall.

Fisher, R. P., Geiselman, R. E., & Amados, M. (1989). Field test of the cognitive interview: Enhancing the recollection of actual victims and witnesses of crime. *Journal of Applied Psychology, 74*, 722–727.

Furlong, N., Lovelace, E., & Lovelace, K. (2000). *Research methods and statistics: An integrated approach.* Orlando, FL: Harcourt Brace.

Gick, M. L., & Holyoak, K. J. (1980). Analogical problem solving. *Cognitive Psychology, 12*, 306–355.

Halonen, J. (2001, August). *Beyond sages & guides: A postmodern teacher's typology*. Harry Kirke Wolfe Lecture at the annual convention of the American Psychological Association, San Francisco, CA.

Heiman, G. W. (2001). *Understanding research methods and statistics: An integrated introduction for psychology* (2nd ed.). Boston: Houghton Mifflin.

Jackson, S. L. (2006). *Research methods and statistics: A critical thinking approach* Second Edition 6. Belmont, CA: Wadsworth/Thomson Learning.

Nisbett, R. E., Krantz, D. H., Jepson, C., & Kunda, Z. (1983). The use of statistical heuristics in everyday inductive reasoning. *Psychological Review, 90,* 339–363.

Novick, L. R. (1988). Analogical transfer, problem similarity, and expertise. *Journal of Experimental Psychology: Learning, Memory, and Cognition, 14,* 510–520.

EXPLANATION OF DOCUMENTS INCLUDED ON THE CD BY ANDREW N. CHRISTOPHER, MARK I. WALTER, ROBERT S. HORTON, AND PAM MAREK

Syllabi.doc

This Word document contains actual syllabi of integrated statistics and research methods courses at Wabash College, Albion College, and St. Michael's College.

Description of Mini-Projects.doc

This Word document contains a description of the miniprojects used at Wabash College in the first semester of the integrated statistics and research methods course sequence.

Implementing a Class Project in Research Methods.doc

This Word document contains information about a project used in a one-semester research methods course (which had statistics as a prerequisite). For teachers at schools without an integrated stats/research methods course, this project provides a means of integrating the two topics in a research methods course.

Survey dataset #1 directions and items.doc
Survey dataset #1.sav

This Word document contains detailed instructions written for an undergraduate research seminar and was to be followed when using the accompanying SPSS file. The spreadsheet has the "raw" data; that is, no reverse-scoring has been done. These documents can be used to teach a variety of topics, such as survey design and nonexperimental data analyses.

Scales Lab Assignment Directions and Items Used.doc
Scales lab assignment dataset.sav

This Word document contains detailed instructions written for students to gain experience working with data derived from psychological scales. Raw data are entered in the accompanying SPSS 12.0 datafile, and students must make all necessary transformations before conducting potential analyses.

Correlation & Regression Class Activity.doc

This Word document contains a hands-on activity for students in which they can devise a hypothesis, operationalize variables, collect data, analyze data by hand, construct a scatterplot, and interpret their results. The activity extends into simple bivariate regression, with analyses linked to SPSS.

Lab Assignment for Dependent/Paired Samples *t*- tests on SPSS.doc
Lab Assignment for Dependent/Paired Samples *t*-tests on SPSS.sav

This Word document contains a hands-on activity for students in which they can create data to use in a two-group repeated measures design, with resultant use of the dependent samples *t*-test. The SPSS 12.0 spreadsheet contains actual class data from this activity from the Spring 2004 semester. Teachers may use these data, or can use the spreadsheet as a template to enter data from their own classes.

One-way repeated measures design stimulus materials.doc
One-way repeated measures SPSS dataset.sav

This Word document contains part of the stimulus materials from a student's class project at Albion College that exemplifies a one-way repeated measures design. The SPSS 12.0 file contains data from 5 respondents and can be used to work a one-way repeated measures ANOVA in class and/or to have students use SPSS with this experimental design.

Basics of Factorial Designs.doc

This Word document contains explanations of and opportunities for students to practice with factorial terminology such as "main effect," "interaction," "marginal means," and "cell means." After an initial lecture on factorial designs, we have found these handouts/possible assignments to be a good way to reinforce the basic lecture material.

Homework-Lab on Factorial Designs.doc
Homework-Lab on Factorial Designs #1.sav
Homework-Lab on Factorial Designs #2.sav
Homework-Lab on Factorial Designs.spo

These four documents can be used to help students gain more experience with factorial designs. The Word document contains two problems that can be used in variety of ways. The first SPSS 12.0 dataset contains the data corresponding to the first problem in the Word document. Although the problem does not require students to work the problem in SPSS, the dataset is here for your convenience. The second SPSS dataset pertains to the second problem on the Word document. The SPSS printout was derived from the second dataset and is required to answer the questions posed in the second question in the Word document.

Chapter 14

Integrating Computer Applications in Statistics and Research Methods

Bernard C. Beins and Ann Lynn
Ithaca College

> *"Technology changes how we teach as well as creating demands for teaching new* content."
>
> —Moore, 1997, p. 3

It is no understatement to assert that, in higher education, computers are ubiquitous. Current generations of students (and even some faculty) probably have no experience of life before computers. When personal computers began to appear in higher education, there were promises of a vast revolution in the way faculty teach and students learn. For instance, Collyer (1984) identified the use of computers to teach thinking rather than simply as powerful computational devices. (He also presaged contemporary teachers [e.g., Daniel, 2004] in noting that it is altogether too easy to use computers badly in teaching.) Beginning in the early 1980s, psychologists began experimenting with this new technology in the classroom. The most prevalent use focused on data collection and statistical analysis (Beins, 1989; Couch & Stoloff, 1989), although there has been no shortage of other suggestions on how to implement computer applications in the classroom (e.g., Ralston & Beins, 1996).

Regarding the use of pedagogical innovations, the extent to which varied approaches to teaching lead to differential student learning is unclear. Historically, researchers have argued that new models of the classroom are no more effective than traditional approaches (e.g., Eglash, 1954; Husband, 1949), nor does it appear that new models of teaching systematically relieve students of their ignorance compared to more traditional modes (Brothen & Wambach, 2001; DeBord, Aruguete, & Muhlig, 2004). Nonetheless, computer use may facilitate student involvement in learning, just as discussion engages students more than listening passively to lectures (Buxton, 1949; Johnson, Zlotlow, Berger, & Croft, 1975; Kirschenbaum & Riechmann, 1975). Perhaps most importantly, the use of computers in statistics and research courses models the behavior of psychologists in their professional activity.

As teachers, we have reached this degree of computer use in our classes through gradual evolution. The initial use of computers, involving self-generated BASIC programs (e.g., Beins, 1984a, 1984b) gave way to more commercially produced software, like Minitab and subsequently SPSS, and other demonstration or experimental software (e.g., Beins, 1990, 1993a). With each new development, our ultimate educational goals tend to remain the same, but we expect our students to engage in different types of activity to reach those goals.

With the advent of the Internet, the availability of resources to facilitate faculty goals has burgeoned, and authoring software for Web pages has recreated the opportunity for individually developed, pedagogically oriented Web pages. In addition, the cost of professional analytical software such as SPSS has dropped, making it accessible to more institutions and students. In this chapter, we identify the ways that one can use computers to good effect in statistics and in research methods classes. Many of these uses reflect the activity of contemporary psychologists. Some are simply updated ways of engaging students in the learning environment to which they have become accustomed.

STATISTICS

Why Technology?

Technology is seductive. When presented with the opportunity to use a sexy new technology to teach a topic that remains relatively static (such as statistics), it is easy to imagine it will lead to a complete revolution in the course. However, it is important to remember that technology is just a tool that can allow achievement of learning objectives. In statistics, those learning objectives are to improve statistical competence, reasoning, and thinking (delMas, 2002). Statistical competence (Rumsey, 2002) refers to a basic foundation in statistical terms, ideas, and techniques as well as interpretation and communication of results. These basic skills and knowledge are the prerequisites for statistical reasoning and thinking. Statistical reasoning (Garfield, 2002) is the ability to explain why or how a result is produced and why a conclusion is (or is not) justified. Statistical thinking (Chance, 2002) is the ability to critique and evaluate conclusions based on statistics and to generalize basic skills to new research situations and to real-world problems. Teachers can reach these objectives without technology, although well-used technology should improve students' learning outcome in these three areas.

Choosing the Right Technology for Introductory Statistics

When selecting a tool to do a job, a carpenter must be aware of the end result and choose the tool that will most efficiently achieve that result. When choosing a technological tool, teachers of statistics must be aware of the goals of the course in statistics and choose the technology that will most efficiently achieve those goals. The statistics in psychology course at our institution emphasizes the development of basic statistical literacy and competence. In the course, students learn statistical reasoning and thinking, but the students develop these skills more fully in our research methods course and in a required three-semester research practicum experience. Because learning basic statistical techniques is a part of developing competence, statistical analysis software is the technological tool that we believe most efficiently helps achieve this goal.

The choice of analytical software depends on the context within which you expect students to use their statistical skills and the resources available. We chose to use SPSS (rather than Excel or a student-oriented program) because it best meets the needs of the students who take the course (see Warner & Meehan, 2001, for a discussion of the relative merits of SPSS and Excel in the teaching of statistics). Specifically, our psychology program has a strong research emphasis, and the departmental faculty expect students to leave the statistics course with the skills to conduct scientific data analyses for and with faculty in the department. We have a site license for SPSS, and our faculty laboratories and many campus computer labs have SPSS. In addition, our statistics and research methods courses take place in a computer classroom with SPSS loaded on each machine. This embarrassment of riches is not necessary to integrate SPSS into the introductory statistics course. Many statistics for psychology texts include a copy of the software for student use.

Balancing Content, Pedagogy, and Technology

One of the traps that even experienced researchers fall into is to conduct more sophisticated statistical analyses than they can understand and interpret. The Task Force on Statistical Inference (Wilkinson & Task Force on Statistical Inference, 1999) recognized that researchers can sometimes fall prey to using complex and inappropriate statistics. As such, the task force cautioned against reporting statistics without understanding how they are computed or what they mean. The Task Force also decried the reliance on results without considering their reasonableness or independent computational verification. The goal of a successful integration of analytical software into an introductory statistics course must be to insure that students fully understand the analyses they conduct and can explain the results. In our experience, successful integration of analytical software is best accomplished by balancing the skills of choosing, generating, and interpreting basic statistics.

To choose the correct analysis, students need to develop a framework for the statistical tools taught in the course. This conceptual framework includes an understanding of the relationships among the analyses and the circumstances under which you use each analysis (Alacaci, 2004; Ware & Chastain, 1991). We teach students to choose the correct test by identifying (a) the variables in the research question; (b) the type of measurement (categorical or continuous) used in the operational definition of each variable; (c) whether the research goal is to describe the research sample (descriptive statistics) or draw a conclusion about the world in general (inferential statistics); and (d) whether the research question refers to finding group differences or relationships (see Fig. 14.1).

Although analytical software can easily generate statistics, we strongly believe that doing basic statistical analyses by hand helps students appreciate the mathematical logic behind the statistics, helps demystify the computations of the computer, and allows the students to check the reasonableness of the output generated by the computer. For

If you have only **1 Variable**

1. Variable is discrete:	Chi-square goodness of fit
2. Variable is continuous:	
a. You know population mean and SD:	*z* test
b. You know population mean:	Single sample *t* test

If you have **2 Variables**

1. Both variables are discrete:	Chi-square test of independence
2. Both variables are continuous:	
a. You want to find the relationship:	Inferences on correlation
b. You want to find difference/influence:	Dependent samples *t* test (a.k.a. paired samples, correlated samples)
c. You want an equation to predict DV:	Regression
3. One variable is discrete and the other is continuous:	
a. The discrete variable has 2 levels (groups):	Independent samples *t* test
b. The discrete variable has 3 or more levels:	One way ANOVA

Figure 14.1. Using scales of measurement to choose the test.

example, calculating the three types of *t* tests by hand reinforces the fact that one compares a sample to a known value, another compares two samples, and the third compares the difference between related samples. The formulas are nothing more than a symbolic representation of these descriptions. These differences are not as apparent when conducting the analyses on the computer. Getting their hands dirty with hand calculations also gives students expectations for the values they should expect to obtain for various descriptive and inferential statistics. These expectations help them give computer output and their interpretations of the output a plausibility check. For example, a student who has conducted one-way ANOVA by hand is likely to recognize that an *F* ratio that is less than 2.0 is probably not going to reach statistical significance, no matter who wrote that conclusion. When students use statistical tables, they get a sense of the values that are likely significant at various degrees of freedom. Although you could eventually figure this out given different computer output, it is clearer in a hand analysis.

Only after students practice analyses by hand do we teach them to conduct the analyses on the computer and read the output. In addition to speeding the calculation of statistics, the computer is useful because it allows for the analysis of more realistic data sets. Because we use hand analyses as a pedagogical tool, we typically present students with tiny data sets (5 to 10 cases per cell). Of necessity, when the results of these analyses are statistically significant, the effect sizes are unrealistically large. Using realistic data sets with larger sample sizes allows students to not only see how real data look, but also to learn that statistical significance and practical significance (effect size) are not always the same. However, it is a mistake to rely on large, existing data sets in a statistics course. Even in this era of computer-aided data collection, many of our data are still hand entered, and it is not always clear to students how to format data for analyses. So we teach students how to enter data into the computer and require them to enter at least one small data set for each statistical analysis they learn.

We use the framework previously described for selecting a test to teach data entry. The data entry mantra in the course is that every participant is a line, every variable is a column, and the categories of discrete variables are coded. Although correct data entry may seem simple, one of the authors (AL) has had many experiences of being asked to consult on professional research projects only to discover that the researchers had entered the data so incorrectly that meaningful results could not be obtained without reformatting the data. With computers, it is garbage in, garbage out (GIGO).

Finally, we emphasize communication of the results of both hand and computer results. Students learn to present the results of their analyses professionally. Practicing professional presentation of the results reinforces the idea that statistics are a tool for answering interesting questions about psychology. It also teaches a practical skill students can use after the statistics course. Writing about statistics also reinforces students' understanding of the analyses (Beins, 1993b; see also Schmidt & Dunn, chap. 17, this volume).

The most effective learning in statistics occurs when the content, pedagogy, and technology work together in a seamless and balanced way. The key to this integration is to inhibit the tendency for technology to be the focus of the course rather than to use it as a pedagogical and practical tool.

RESEARCH METHODS

Caveats About Using Technology

It is indisputable that no area is so interesting that a clever academic cannot make it boring. There is a corollary to this maxim: No technology is so powerful that a clever teacher cannot make it dreadful. Instructors need to think about the implications of the way they use technology. In its most neutral incarnation, technology reproduces the static handouts that have graced classrooms for generations of students. Unfortunately, some computerized versions of those handouts may block learning. As Daniel (2004) pointed out, the inclusion of sound and animation in PowerPoint slide transitions may engender more irritation by students than learning. If incorporating interesting asides in a lecture diminishes learning (Harp & Maslich, 2005), one would expect even more deterioration if these asides also annoy students.

So the question arises as to what would constitute good use of technology in the research methods course. One answer is to use a Web interface for the research methods syllabus. A good use of Web technology should involve many of the same characteristics that students associate with good faculty Web pages: contact information, course information, and content information (Palmiter & Renjilian, 2003). In addition, appearance can enhance what visitors gain by visiting the Web site (Plous, 2000). In terms of student responses, Sherman (1998) noted that students are negatively affected when an Internet site's format overwhelms the content.

In essence, all materials for the Research Methods course are online, although the class is classroom based. We are fortunate to be able to meet in a computer classroom, so all the materials are accessible all the time and for all students. Because of my (BB) reliance on computerized presentation of materials, it is important for me to maintain the Web site and all materials with both pedagogy and usability in mind. The examples provided here reflect the decisions I have made regarding what I hope students learn and how best to present the online material. My online syllabus (Beins, 2004), with most of the activities and much information about the course, is available via the Office of Teaching Resources in Psychology (OTRP) and can be downloaded from that site (http://www.teachpsych.org). Its general appearance is evident in Figure 14.2, and it appears in full on the CD accompanying this book. The syllabus includes the various applications of the Internet in my course, including homework assignments, data collection exercises, and online tests.

Using Forms

Creating forms for the Internet is a useful skill to develop. Forms allow you to present information you want to disseminate and, most usefully, to collect information or feedback from students. You can gather such information from students in traditional ways; students can write assignments out by hand and submit them in person, or they can word process their homework. Most students are also aware of how to send attachments to their instructors via e-mail. As such, they can complete a homework assignment in a word processing program and turn it in electronically as an attachment. This approach is easy for students and is a useful technique for collecting homework. Surprisingly, a significant number of students still prefer to hand in their homework on paper when given the choice, even though they create their homework using word processing software.

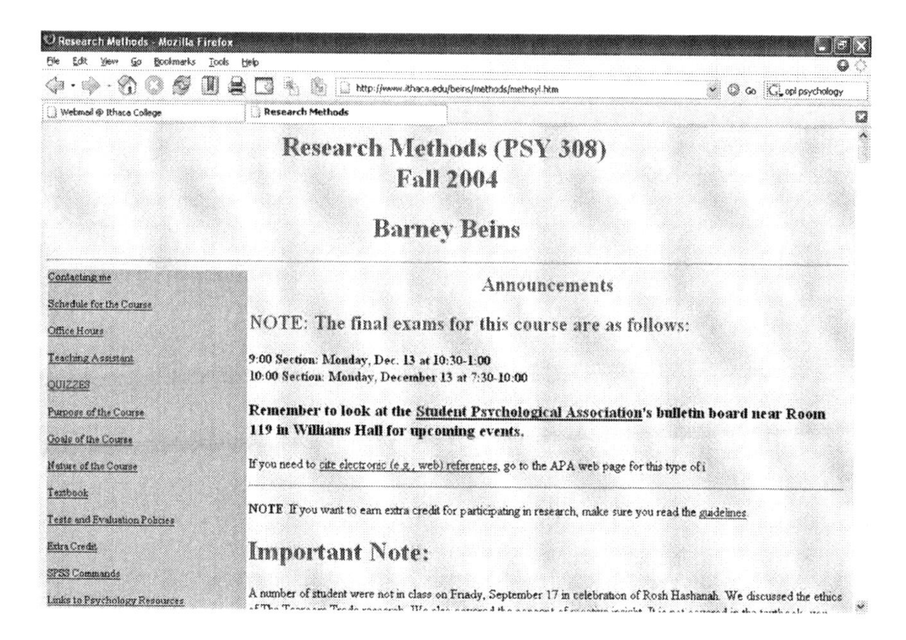

Figure 14.2. Electronically available syllabus for the Research Methods class, with an index (on left) of contents of the syllablus.

One of the problems that arises when students use standard word processing to do their homework is that the format of their answers can be quite variable. Most instructors probably prefer standardization of homework formatting when possible. Standardization makes grading easier and quicker. To ensure receipt of a homework assignment in a consistent format, it is useful to create forms for the Internet that students can use to complete and submit the assignments.

Various programs like *Dreamweaver*® and *Front Page*® allow for easy creation of forms even if the user does not know hypertext markup language (HTML). It is useful to know HTML because you can sometimes tailor the format of your page more easily using HTML code. But for most forms you are likely to create, there is no need to know HTML. As with any new technological approach, there is a learning curve for becoming comfortable with creating forms, but the task of creating these Web pages is not onerous. The advantage of using this software is that you see immediately the format of the page in a What You See Is What You Get (WYSIWYG) format.

Developing forms requires formatting of the Web page, followed by a processing script that tells the computer what to do with the information that the respondent provides. Creating the form is relatively easy. Writing the processing script is slightly more demanding because the technical formatting of the script is very sensitive to departures from appropriate style. Your institution may offer tutorials or workshops on this topic. In this chapter, the discussion only outlines the creation of the forms for accepting homework, for data collection, and for online testing, leaving the reader to pursue the details for processing scripts.

The various commercial offerings (e.g., Blackboard, WebCT) can alleviate the need for processing scripts for tests and quizzes. One drawback to these services, however, is a degree of incompatibility when changing from one to another. The use of self-generated forms and pages obviates some of the incompatibility issues.

Homework

One example in my research course involves analysis of research with an assessment of whether it shows the characteristics of scientific research. Submission by students of this homework is through a Web-based form. Francis Galton (1872) examined the efficacy of prayer and discovered that Britain's royal family did not live longer than other groups (clergy, doctors, and lawyers), even though the British populace prayed regularly for the royal family. There are enough flaws in the study and its conclusion to generate lively discussion. Figure 14.3 shows the form as it appears on the Web; it also appears on the CD accompanying this book.

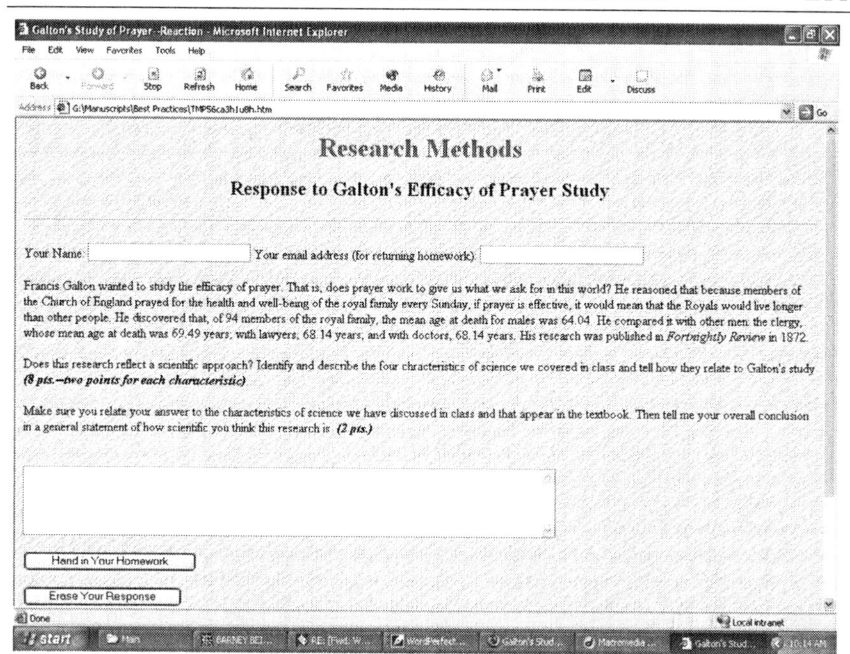

Galton's Study of Prayer -Reaction - Microsoft Internet Explorer

Research Methods

Response to Galton's Efficacy of Prayer Study

Your Name: Your email address (for returning homework):

Francis Galton wanted to study the efficacy of prayer. That is, does prayer work to give us what we ask for in this world? He reasoned that because members of the Church of England prayed for the health and well-being of the royal family every Sunday, if prayer is effective, it would mean that the Royals would live longer than other people. He discovered that, of 94 members of the royal family, the mean age at death for males was 64.04. He compared it with other men: the clergy, whose mean age at death was 69.49 years; with lawyers, 68.14 years; and with doctors, 68.14 years. His research was published in *Fortnightly Review* in 1872.

Does this research reflect a scientific approach? Identify and describe the four chracteristics of science we covered in class and tell how they relate to Galton's study ***(8 pts.--two points for each characteristic)***

Make sure you relate your answer to the characteristics of science we have discussed in class and that appear in the textbook. Then tell me your overall conclusion in a general statement of how scientific you think this research is. ***(2 pts.)***

Hand in Your Homework

Erase Your Response

Figure 14.3. Example of a Web form for student homework.

This assignment is not complex in the sense of having multiple components. For more multifaceted homework, though, it is convenient to include several text boxes and to format the output so that each paragraph or section of an answer is separated in the output for convenient viewing and grading.

The processing script for this homework form directs the computer to e-mail the answer to the instructor. Converting the e-mail to a Word document permits use of the Comment feature to insert remarks and the student's score. For instructors who can type faster than they can write longhand, the grading process is quicker for assignments that call for feedback, most of which do. Further, the instructor can e-mail the homework back to the student at any time.

Students appear to appreciate receiving homework back in this format. They get their homework back in a more timely fashion; they do not have to wait for a class meeting to see it. In addition, comments can be more detailed and provide students with more useful feedback.

Data Collection in Class

As numerous psychologists have asserted, students enjoy providing data about themselves and, as a result of providing this data, psychological questions and data analysis have more meaning for them (e.g., Day, Marshall, & Rubin, 1998; Rowland, Kaariainen, & Houtsmuller, 2000). With this fact in mind, many instructors try to incorporate meaningful data collection exercises in their statistics and research courses.

In my Research Methods class, students learn about various research methodologies by acting as participants. Forms on the Web provide a convenient means of doing so. The approach is fairly simple and does not require knowledge of programming language. Consequently, creating the forms is quick, although there is loss of flexibility that more sophisticated and technical programming allows in some matters, like automatic assignment of students to different experimental or control groups.

The example here involves correlational research in which students learn about the concept of the *body mass index* (BMI). The form appears in Figure 14.4 and on the accompanying CD. Students look up their BMI on the Shape Up America Web site http://www.shapeup.org/bodylab/frmst.htm), measure their waist and hips with a tape measure that I provide, give a self-rating of the healthfulness of their lifestyle, and indicate their sex.

BMI is associated with relative risk of cardiovascular disease, with extremely high and low values of the BMI being associated with higher relative risk of death. BMI is not a perfect predictor of one's risk for dis-

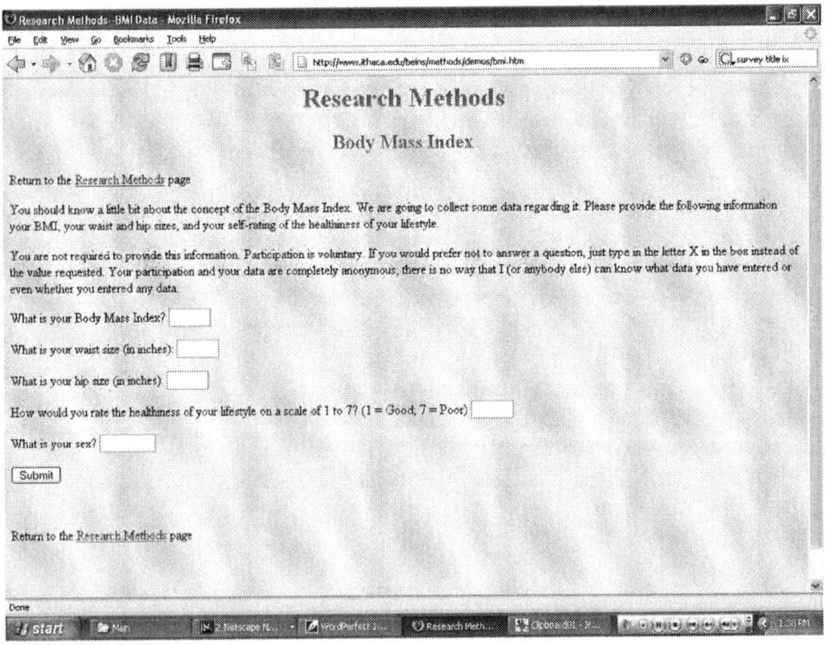

Figure 14.4. A Web form for data collection in a correlational study for which students provide data.

ease, but research has clearly shown its utility (Calle, Thun, Petrelli, Rodriguez, & Heath, 1999).

One might expect that unhealthy lifestyles would generally be associated with high BMI and with greater risk of death. On that basis, greater healthfulness of lifestyle should correlate with more moderate levels of the BMI. This data collection exercise investigates the potential relation between these variables. At first, students believe that the self-report of lifestyle should correlate with BMI, but the relation fails to materialize.

When students complete the form, I send data from each person to a file that can be imported easily into SPSS. Combined data from different classes across semesters totaling 223 students produced a correlation about as close to zero as it can be, $r(221) = .01, p = .88$. Students may be tempted to focus on the limitations of the BMI, but there is convincing evidence that lifestyle issues (e.g., smoking, lack of exercise) with a negative impact on health are associated with high levels of BMI (Smalley, Wittler, & Oliverson, 2004). This result leads to a discussion of how to ask better questions to see whether lifestyle is associated with BMI. The question used here intentionally does not provide useful information on lifestyle, so students can pursue the issue in class discussion.

Another aspect of the data that students can investigate is the waist-to-hip (WHR) ratio. Data reported by Bamia, Trichopoulou, Lenas, and Trichopoulos (2004) revealed that among smokers, BMI was associated with a larger WHR ratio. The data set indicates such a significant positive correlation for the entire set of students, $r(221) = .24, p < .001$. In addition to this type of data collection, a new and growing set of experiments, demonstrations, and simulations will help teachers incorporate data collection into their classes. The Online Psychological Laboratory (OPL; http://opl.apa.org/Main.aspx), initially developed by Ken McGraw at the University of Mississippi, will be housed on the Web site of the American Psychological Association. It will provide free access to experiments in which students add to an existing data set. Instructors will be able to download data for the exercises for appropriate statistical analysis. This National Science Foundation-funded initiative will allow greater sharing of resources in research design and analysis.

Testing Students Online

Students take their tests in the Research Methods class online while in the classroom. (Students have the option to take tests on paper. Over the past 3 years, one student availed herself of that opportunity, and it was for only one test.) For multiple-choice questions, students select the best option; they also have the chance to explain their answer in a text box immediately below the question. If they think there are multiple correct answers or no correct answer, they can adduce an argument explicating their answer. This strategy of providing students with the option to comment on multiple-choice questions changes student scores minimally, but it appears to lessen student anxiety (Nield & Wintre, 1986).

As with any other form, tests are fairly easy to create for online use. The computer e-mails the students' responses to me; these responses are converted to Word format for grading. For grading the essay portion of a test, the Comments feature allows for easy and extensive feedback to students about where they lost points and why. The advantage of this format is that the tests can be quicker to score and return than if students write their essay answers longhand. For teachers who make comments and who type well, it is very quick and easy to make comments for the students' benefit. Another practical benefit is that the instructor can return the test as soon as it is graded. One application of this advantage is that the instructor can give students the opportunity to look over the questions and raise issues during the subsequent class period. When I return tests to my students, I e-mail them with the scored test attached. As a result, we both retain a permanent copy. Any uncertainty about a student's final accumulation of points is easy to resolve.

A further possible advantage is that it may be harder for students to cheat opportunistically on the test. That is, if they do not coordinate a cheating strategy but let their eyes wander to the tests of other students, it can be hard for them to tell which question another student is reading because, as students progress through the test, they scroll down to see a new question. Students working at different speeds will be looking at different questions.

Just as advantages accrue with online testing, there are some limitations as well. For example, this type of testing requires that each student be at a computer. This disadvantage can be overcome with remote testing, which has its own demands. In addition, online tests using forms can be time-consuming to create because each multiple-choice question needs four or five response buttons; inserting them can be tedious. (Commercial software for course organization, like WebCT or Blackboard makes such testing easier.)

A final disadvantage of this testing format is that the instructor is chained to the computer for grading. Reading large amounts of text on a screen is not particularly enjoyable; this testing requires it. It is always possible to print the test answers out, but such a strategy defeats some of the advantages of online testing.

Overall Advantages and Disadvantages

Advantages. Once an instructor is conversant with constructing and formatting basic Web pages and forms, it is easy to move one's class online. Even if each student is not seated at a computer, the instructor in a wired classroom can project the relevant pages for everyone in the class to see. Creating a Web page usually takes about as much time as it would to word process a traditional handout. With the word processor's cut-and-paste feature, incorporating material onto a Web page speeds the task significantly.

Updating Web pages is also technically quite easy, although the related reality is that maintaining and updating Web site requires constant vigilance. Because making a Web page is no more involved than most word processing, however, revisions need not be onerous. In fact, Word and Word Perfect users can save their text in HTML. This approach leads to text-heavy pages, but one can insert pictures if desired.

For teachers who are interested in learning to develop their Web pages, standard Web authoring software like *Dreamweaver* or *Front Page* will allow easy design of attractive Web pages. It helps at times to know HTML. Free, useful tutorials are available on the Web (e.g., at http://www.w3schools.com/html or at http://www.davesite.com/webstation/html/). More advanced but understandable details are also free on the Internet (e.g., http://www.htmlgoodies.com/).

Regarding content, there is no limit to how much you can provide to your students. There are numerous sources of material on the Web to which you can send your students. So, in addition to posting your own creations, you can give students the benefit of the work of others. A further advantage of keeping materials on the Web is that there may be less chance of losing them. You put them on the Web rather than in a pile of papers that can get lost on your desk or filed in an inappropriate folder. Similarly, students are not able to lose the material you give them. It is available on the Web all the time. Related to this advantage, material on the Internet does not use paper. So not only do you save this resource, but you do not lose time sending material someplace to be copied.

Disadvantages (and Ways to Overcome Them). Even though it is easy to make changes in your Web pages in order to keep them up to date, it takes time. If you fail to keep your Web site current, your students will experience frustration when they try to use broken link. If you plan your class around some resource on the Web, an inactive links spoils everybody's day. Advance planning cures this situation.

I have identified another problem that I call "resource creep." The amount of material I have developed and the number of external resources I have found accumulate so that it is impossible for me to incorporate it all into my class. Failure to plan which materials to use and to incorporate them seamlessly into the classroom induces an appearance of chaos. I need periodic reminders that my students have limited time and motivation, so I have to restrict the amount of material that I assign. I also realize that I do not need to access links on my Web site simply because they are there.

A final potential problem is complacency in preparing for classes. It is too easy to assume that you have prepared for class just because you have access to materials for the day's session. This problem is not unique to the Internet, but it might be more likely to occur compared to the situation where you have to write out your notes prior to class. Web pages, by virtue of their being on the Internet, take on a reality of their own. So you can mistakenly conclude that you remember the content of the Web pages and that you know how you intend to use the material simply because it is there. My solution to this problem is to spend an hour prior to each class, preparing what I want to say and how I am going to say it. The Web material merely facilitates what emerges from my planning session.

ASSESSING STUDENT LEARNING

Students completing an introductory statistics in psychology course should have (a) proficiency at the use of descriptive and basic inferential statistics, (b) understand the foundations of statistical inference, (c) select and conduct appropriate statistical analyses for simple projects, and (d) professionally communicate the results of their analyses (Halonen et al., 2003). Although basic proficiency using analytical software is not among these outcomes, we believe it both facilitates learning and models the behavior of professionals. It adds to the practical value of an education in psychology.

We have advocated teaching students to analyze research questions to determine the variables involved in the question, the type of measurement employed, and the purpose of the research. This structure aids students in selecting the appropriate analysis and correctly hand-entering data. Requiring students to conduct analyses by hand promotes their understanding of the differences among inferential tests and enables them to gain confidence in analytical skills. It also may help students better appreciate their roles as statisticians and reduce their tendency to take on faith the results produced by the computer. Although we do not have data to test this hypothesis, we have observed many students realize they had selected the wrong analysis when the SPSS output did not look "right." Their expectations about the results came from experience conducting hand analyses.

We use SPSS as a tool to aid in our final assessment of proficiency in introductory statistics. The final exam requires students to use SPSS to answer several basic research questions about an existing, real, data set we provide. In addition, we require them to generate and answer their own research question based on the data. These tasks address the skills we expect students to master in the course and provide a foundation for the more independent statistical reasoning and thinking they will encounter in subsequent courses.

As with statistics, initial assessment of proficiency in research methods involves testing on course content and evaluation of proficiency in applied homework problems. These traditional assessments require some integration of concepts and applications of statistics in the research context.

In the long term, assessment is more oriented toward student performance in their research activities. A survey of our alumni revealed considerable satisfaction with their research experience; further, the number of successful student research projects leading to conference presentations gives the department a picture of student research competence. Given students' enjoyment of research, their belief that it has been useful for them, their strong record of conference presentations, and finally, their success in being accepted into graduate programs, the appropriate conclusion is that our empirical sequence of Statistics and Research Methods is meeting the needs of the students and of the department.

CONCLUDING THOUGHTS

Over the years, psychologists have adopted varied technologies for teaching, including radio (Gaskill, 1933; Snyder, Greer, & Snyder, 1968), commercial television (Husband, 1954), and telephone (Cutler, McKeachie, & McNeil, 1958). The result is sometimes better test performance by students, but new and exciting technologies usually fail to live up to their billing as harbingers of educational revolution.

Nonetheless, computer applications can make a difference in the classroom. First of all, in statistics, computers relieve students of the tedium of onerous hand calculations and allow instructors to teach with realistic, interesting data sets. Exempting the student from time-consuming hand calculations frees up time for other learning, like understanding why small *p* values are not the Holy Grail of research or how to communicate statistical and research results in comprehensible language. Further, because psychologists use computers for data collection and statistical analysis and because contemporary students have adapted their learning to the use of computers, it is incumbent on teachers to model the behavior of psychologists and to encourage learning through a medium with which students are comfortable.

Developing new technologies forces the teacher to attend to pedagogy. Nonetheless, as Wolfe (1895) sagely pointed out, in teaching about research, teachers have to be willing to devote time to their students' needs. He asserted that "if instructors in psychology are unwilling to do this kind of work, we must wait until another species of instructor can be evolved" (p. 385). Perhaps the greatest effect of the new technology is that learning the new technologies energizes teachers and promotes their revolutionary zeal for the benefit of the learner.

REFERENCES

Alacaci, C. (2004). Inferential statistics: Understanding expert knowledge and its implications for statistics education. *Journal of Statistics Education, 12*. Retrieved May 15, 2005 from www.amstat.org.publications/jse/v12n2/alacaci.html

Bamia, C., Trichopoulou, A., Lenas, D., & Trichopolos, D. (2004). Tobacco smoking in relation to body fat mass and distribution in a general population sample. *International Journal of Obesity, 28*, 1091–1096.

Beins, B. C. (1984a, July). *A hands-on approach to experimental psychology. Computing in undergraduate psychology*. Presentation at Practicum on Computing in Psychology, Gettysburg College, Gettysburg, PA.

Beins, B.C. (1984b, July). *Demonstrations of complex cognitive phenomena*. Presentation at Practicum on Computing in Psychology, Gettysburg College, Gettysburg, PA.
Beins, B. C. (1989). A survey of computer use reported in *Teaching of Psychology*: 1974–1988. *Teaching of Psychology, 16*, 143–145.
Beins, B. C. (1990). Computer software for introductory psychology courses. [Review]. *Contemporary Psychology, 35*, 421–427.
Beins, B. C. (1993a). Perception on computer: A diverse approach [Review]. *Contemporary Psychology, 38*, 100–101.
Beins, B. C. (1993b). Writing assignments in statistics class encourage students to learn interpretation. *Teaching of Psychology, 20*, 161–164.
Beins, B. C. (2004). *Research methods syllabus*. Available at Office of Teaching Resources in Psychology Web site: http://www.lemoyne.edu/OTRP/projectsyllabus.html
Brothen, T., & Wambach, C. (2001). Effective student use of computerized quizzes. *Teaching of Psychology, 28*, 292–294.
Buxton, C. E. (1949). The pros and cons of training for college teachers of psychology. *American Psychologist, 4*, 414–417.
Calle, E. E., Thun, M. J., Petrelli, J. M., Rodriguez, C., & Heath, C. W. (1999). Body-mass index and mortality in a prospective cohort of U.S. adults. *New England Journal of Medicine, 341*, 1097–1105.
Chance, B. L. (2002). Components of statistical thinking and implications for instruction and assessment. *Journal of Statistics Education, 10*. Retrieved May 15, 2005 from www.amstat.org.publications/jse/v10n3/chance.html
Collyer, C. E. (1984). Using computers in the teaching of psychology: Five things that seem to work. *Teaching of Psychology*, 11, 206–209.
Couch, J. V., & Stoloff, M. L. (1989). A national survey of microcomputer use by academic psychologists. *Teaching of Psychology, 16*, 145–147.
Cutler, R. L., McKeachie, W. J., & McNeil, E. B. (1958). Teaching psychology by telephone. *American Psychologist, 13*, 551–552.
Daniel, D. (2004, January). *How to use technology to ruin a perfectly good lecture*. Presented at the National Institute on the Teaching of Psychology, St. Petersburg Beach, FL.
Day, H. D., Marshall, D. D., & Rubin, L. J. (1998). Statistics lessons from the study of mate selection. *Teaching of Psychology, 25*, 221–224.
DeBord, K. A., Aruguete, M. S., & Muhlig, J. (2004). Are computer-assisted teaching methods effective? *Teaching of Psychology, 31*, 65–68.
delMas, R. C. (2002). Statistical literacy, reasoning, and learning: A commentary. *Journal of Statistics Education, 10*. Retrieved May 15, 2005 from http://www.amstat.org/publications/jse/v10n3/delmas_intro.html
Eglash, A. (1954). A group-discussion method of teaching psychology. *Journal of Educational Psychology, 45*, 257–267.
Galton, F. (1872). Statistical inquiries into the efficacy of prayer. *Fortnightly Review, 12*, 124-135. Retrieved July 1, 2005 from http://www.mugu.com/galton/essays/1870-1879/galton-1872-fort-rev-prayer.pdf
Garfield, J. (2002). The challenge of developing statistical reasoning. *Journal of Statistics Education, 10*. Retrieved May 15, 2005 from www.amstat.org.publications/jse/v10n3/garfield.html
Gaskill, H. V. (1933). Broadcasting versus lecturing in psychology: Preliminary investigation. *Journal of Applied Psychology, 17*, 317–319.
Halonen, J. S., Bosack, T., Clay, S., & McCarthy, M. (with Dunn, D. S., Hill, IV, G. W., McEntarfer, R., Mehrota, C., Nesmith, R., Weaver, K. A., & Whitlock, K.)

(2003). A rubric for learning, teaching and assessing scientific inquiry in psychology. *Teaching of Psychology, 30,* 196–208.

Harp, S. F., & Maslich, A. A. (2005). The consequences of including seductive details during lecture. *Teaching of Psychology, 32,* 100–103.

Husband, R. W. (1949). A statistical comparison of the efficacy of large lecture versus smaller recitation sections upon achievement in general psychology. *American Psychologist, 4,* 216.

Husband, R. W. (1954). Television versus classroom for learning general psychology. *American Psychologist, 9,* 181–183.

Johnson, W. G., Zlotlow, S., Berger, J. L., & Croft, R. G. (1975). A traditional lecture versus a PSI course in personality: Some comparisons. *Teaching of Psychology, 2, 156–158.*

Kirschenbaum, D. S., & Riechmann, S. W. (1975). Learning with gusto in introductory psychology. *Teaching of Psychology, 2,* 72–76.

Moore, D. S. (1997). New pedagogy and new content: The case of statistics. *International Statistical Review, 65,* 123–165.

Nield, A. F., & Wintre, M. G. (1986). Multiple-choice questions with an option to comment: Student attitudes and use. *Teaching of Psychology, 13,* 196–199.

Online Psychological Laboratory. (n.d.). Retrieved May 24, 2005 from http://opl.apa.org/Main.aspx

Palmiter, D., Jr, & Renjilian, D. (2003). Improving your psychology faculty home page: Results of a student–faculty online survey. *Teaching of Psychology 30,* 163–166.

Plous, S. (2000). Tips on creating and maintaining an educational World Wide Web site. *Teaching of Psychology, 27,* 63–70.

Ralston, J. V., & Beins, B. C. (1996, November). Thirteen ways to computerize your course. *APS Observer,* 24–27.

Rowland, D. L., Kaariainen, A., & Houtsmuller, E. (2000). Interactions between physiological and affective arousal: A laboratory exercise for psychology. *Teaching of Psychology, 27,* 34–37.

Rumsey, D. J. (2002). Statistical literacy as a goal for introductory statistics courses. *Journal of Statistics Education, 10.* Retrieved May 15, 2005 from www.amstat.org.publications/jse/v10n3/rumsey2.html

Sherman, R. C. (1998). Using the World Wide Web to teach everyday applications of social psychology. *Teaching of Psychology, 25,* 212–216.

Smalley, S. E., Wittler, R. R., & Oliverson, R. H. (2004). Adolescent assessment of cardiovascular heart disease risk factor attitudes and habits. *Journal of Adolescent Health, 35,* 374–379.

Snyder, W. U., Greer, A. M., & Snyder, J. (1968). An experiment with radio instruction in an introductory psychology course. *Journal of Educational Research, 61,* 291–296.

Ware, M. E., & Chastain, J. D. (1991). Developing selection skills in introductory statistics. *Teaching of Psychology, 18,* 219–222.

Warner, C. B., & Meehan, A. M. (2001). Microsoft Excel as a tool for teaching basic statistics. *Teaching of Psychology, 28,* 295–298.

Wilkinson, L., & Task Force on Statistical Inference. (1999). Statistical methods in psychology journals: Guidelines and explanations. *American Psychologist, 54,* 594–604.

Wolfe, H. K. (1895). The new psychology in undergraduate work. *Psychological Review, 2,* 382–387.

Chapter 15

Through the Curriculum and Beyond: Giving Students Professional Opportunities in Research

Randolph A. Smith
Kennesaw State University

Psychology has documented the evolution of its curriculum through curriculum surveys that began in the 1930s (Henry, 1938). Similarly, study groups and national conferences have convened to study the curriculum periodically from the 1950s on. Statistics and research courses have a long history of inclusion in psychology curricula in the United States, as documented by these curriculum reports. In this chapter, I focus on how the curriculum encourages psychologists to train students in statistics and research methodology. However, the curriculum typically does not focus on having students apply this knowledge as professional psychologists do. The last portion of the chapter focuses on some practices for giving students professional opportunities after their statistics and research training; it is these opportunities that I see as best practices.

National Conferences on the Undergraduate Psychology Curriculum

The first national study group (Buxton et al., 1952) met at Cornell University in the summer of 1951 to develop a set of recommendations for the undergraduate psychology curriculum (Brewer, 1997). The Cornell group's report included the recommendation that departments should teach psychology "as a scientific discipline in the liberal arts tradition" (Brewer, 1997, p. 435). To accomplish this goal, their recommended curriculum included a stand-alone course in statistics. Rather than recommending a separate course in experimental psychology or research methods, the Cornell group advised teaching research methodology by including methods in all substantive courses; they recommended that core courses of statistics, motivation, perception, thinking and language, and ability all should have laboratories included.

A follow-up conference took place at the University of Michigan in 1960. They, too, issued a report on the state of the undergraduate psychology curriculum (McKeachie & Milholland, 1961). The Michigan group did not agree to endorse one curricular model, so they described three possibilities (Brewer, 1997). One of their models—the hourglass model—included a yearlong course in statistics and experimental design.

About a decade later, the American Psychological Association (APA) received a grant from the National Science Foundation to study the undergraduate curriculum (Brewer, 1997). James Kulik of the University of Michigan headed this project, which resulted in a report (Kulik, 1973), but no curriculum recommendations. Perhaps reflecting the spirit of the 1960s, the Kulik report shied away from describing an ideal curriculum and, instead, focused on a national survey of psychology curricula. In 1984, APA commissioned another national curriculum survey, which again resulted in a descriptive report (Scheirer & Rogers, 1985) rather than a layout of an ideal curriculum.

The next curriculum study group grew out of an initiative pushed by the Association of American Colleges to study the history and status of several undergraduate majors, including psychology (Brewer, 1997). McGovern, Furumoto, Halpern, Kimble, and McKeachie (1991) developed eight goals that they hoped would serve as guidelines in developing psychology curricula; one goal involved research methods and statistical skills. McGovern et al. believed "these skills should be fostered in separate courses, developed in laboratory work, and reinforced by the use of critical discussion of research findings and methods in every course" (p. 601). This group developed several possible curriculum models—a generalist model and multiple thematic models (emphases on developmental, biological, and health). All curricular models recommended statistics and methods courses after the introductory psychology course.

Yet another national group to study the psychology curriculum (along with six other topics) convened at the St. Mary's Conference at St. Mary's College of Maryland in 1991. Spearheaded by Charles Brewer of Furman University, the curriculum task force issued a report in the conference proceedings (Brewer et al., 1993). The diverse group of nine individuals in the curriculum task force did not come to a consensus model curriculum; they "concluded that no one curriculum is ideal for every school and every student but that all programs should reflect certain common characteristics and suggestions" (Brewer, 1997, p. 439). Although there was no consensus about an overall curriculum model, the task force did agree about the importance and place of statistics and research methods in the curriculum:

> Usually taken in the second or third year, the crucial methodology courses should cover experimental, correlational, and case study techniques of research, and they should involve firsthand data collection, analysis, and in-

terpretation. Methodology courses should cover statistics, research design, and psychometric methods, and they should be prerequisites for some of the content courses. (Brewer, 1997, p. 439)

The task force's emphasis on the importance of teaching undergraduates a scientific approach to the discipline came through strikingly in The Principles for Quality Undergraduate Psychology Programs (Baum et al., 1993; see also McGovern & Reich, 1996), which the Steering Committee members of the St. Mary's Conference developed. Scattered throughout this document are phrases such as "the curriculum enables students to think scientifically about behavior and mental processes" (p. 19) and "the curriculum is based on clear and rigorous goals .... These include thinking scientifically; understanding the relationships between theories, observations, and conclusions; critically evaluating the empirical support for various theories and findings" (p. 19).

The most recent national group to weigh in on the psychology curriculum was a task force appointed by APA's Board of Educational Affairs to develop a set of goals and learning outcomes for undergraduate psychology majors. In its report (Task Force on Undergraduate Major Competencies, 2002), the task force agreed with previous study groups that defining a national curriculum was impossible; in fact, the report even acknowledged that agreeing on goals and learning outcomes would be difficult across different institutions. Regardless, the task force did come to a consensus on a learning goal involving statistics and research training.

Goal 2. Research Methods in Psychology

According to the Task Force on Undergraduate Psychology Major Competencies (2002),"Students will understand and apply basic research methods in psychology, including research design, data analysis, and interpretation" (p. 8). Although it is clear that most of these national study groups have been hesitant to prescribe a national curriculum for psychology, it is also clear that they have valued and recommended the training of undergraduate psychology majors in statistics and research methods. This recommendation clearly would appear to be a best practice recommendation for psychology departments around the country. As these national groups have issued their reports, curriculum surveys of the discipline allow a look at the degree to which psychology departments have incorporated the national groups' recommendations.

Curriculum Surveys

Various interested parties have periodically conducted surveys of the undergraduate psychology curriculum. The first person to make such a survey may have been Henry (1938). According to Perlman and McCann (1999a), Henry tallied the 25 most frequently listed courses;

statistics and research made that list. Beginning with Sanford and Fleishman (1950), it was a common practice to list the top 30 courses in the curriculum (Perlman & McCann, 1999a). Using this approach, Perlman and McCann (1999a) conducted a large-scale curriculum survey in 1997 and compared their results to the earlier comparable curriculum surveys of Daniel, Dunham, and Morris (1965); Kulik (1973); and Lux and Daniel (1978). Perlman and McCann (1999a) found that the percentage of all departments listing statistics had increased from 36% to 43% to 46% to 48% during the 36-year interval. Because few 2-year colleges offered statistics (14% in 1997), the figure was considerably higher for doctoral universities (72% in 1997) and comprehensive colleges and universities (59% in 1997), but not baccalaureate colleges (47% in 1997). Other surveys at about the same time found figures of 58% (Perlman & McCann, 1999b) and 77% of departments requiring a statistics course (Messer, Griggs, & Jackson, 1999). Possible reasons that that figure is not higher are that some departments may have students complete a statistics course in a mathematics department or that they may subsume statistics within a methodology course (23% of departments, according to Messer et al., 1999).

Perlman and McCann (1999a) found that experimental psychology had hovered around 45% for all types of colleges during the time periods they studied. Again, experimental psychology was rarely offered in 2-year colleges (7% in 1997), so the figure was higher for doctoral universities (57% in 1997), comprehensive colleges and universities (59% in 1997), and baccalaureate colleges (51% in 1997). Although these figures are not as high as one might expect, they may be reduced due the number of schools that have begun to offer a course in research methods: 42% overall (up from 21% in 1975), 58% for doctoral universities (up from 34% in 1975), 49% for comprehensive colleges and universities, and 48% (up from 26% in 1975) for baccalaureate colleges. In addition, a course titled research participation was offered by 28% of all colleges (47% doctoral universities, 34% comprehensive colleges and universities, 30% baccalaureate colleges). Messer et al. (1999) found that 73% of departments required at least one methods course; Perlman and McCann (1999b) found that 40% of their departments required research methods, 38% required experimental psychology, and 7% required experimental design. It appears to be a safe assumption that the vast majority of colleges and universities offer their psychology majors some type of research training. Thus, it appears that the best practice recommendation of offering statistics and research training to psychology majors has indeed been put into practice.

Why Teach Statistics and Research?

Given the strong emphasis on teaching statistics and research methodology to psychology undergraduates, an obvious question that arises is "Why teach these topics?" It is clear from the preceding curriculum review that psychology faculty consider these topics to form part of the core knowledge of the discipline. It seems that most, if not all, psychology faculty advocate that psychology is a science. At the same time, students in psychology classes often expect the major to involve little but applied, helping-oriented types of courses. These disparate expectations often lead to misunderstandings between the two groups.

A variety of authors in this volume have enumerated the most common reasons for requiring students to take statistics and research methods courses beyond the simple historical precedent just reviewed. Paul Smith (chap. 5, this volume) listed the reasons of students being aware that psychological knowledge has empirical support, learning to think critically, and developing research skills. Similarly, Ault, Munger, Tonidandel, Barton, and Multhaup (chap. 9, this volume) listed four goals of Davidson's College methodology requirement: helping students become (a) contributors to psychological knowledge, (b) informed consumers of research findings, (c) stronger applicants for graduate study, and (d) satisfied partners in a learning relationship with a faculty member.

A common outgrowth of requiring students to take statistics and research courses is the additional requirement that they complete a research project of some type. It is difficult to know exactly how widespread this type of requirement is. Kierniesky (2005) found that a variety of research-related activities had increased since his earlier survey of small psychology departments on the same topic (Kierniesky, 1984). For example, in the 20-year time interval, he found that the average number of projects per department increased threefold, presentations at professional meetings increased about 450%, and presentations at undergraduate meetings increased over 400%. Undergraduate publications increased a more modest 30%. Kierniesky (2005) found that 63.5% of the departments required an undergraduate research project, compared to only 22.9% two decades earlier. Perlman and McCann (2005) conducted a curriculum study of 500 colleges and found that 79% of the schools required at least one course with "research opportunities" (p. 9). However, only 26% of the respondents said that they required an undergraduate thesis or "substantive project (senior experience)" (p. 10). It is possible that Perlman and McCann's options were higher level requirements than Kierniesky's choice, which might account for the sizable difference in their figures.

Many authors in this volume have mentioned an additional justification for going beyond the basic requirements of statistics and methodology courses and requiring an actual student research project. Instead of merely reading about the steps involved in doing research, students get hands-on experience by completing a research project (Ault et al., chap 9, this volume). By engaging in the research process, students derive the benefits of active learning (Davis, chap. 2, this volume) and avoid the difficulty of transferring book knowledge to real life (see Christopher, Walter, Horton, & Marek, chap. 13, this volume).

Growth of Psychology Student Research Conferences

Psychology student research conferences typically exist to give psychology undergraduates a forum for making research presentations in an environment modeled after professional conferences for faculty. Typically, such conferences take place over 1 or 2 days and feature many student paper (or poster) presentations, an invited speaker, and a group meal or social event (Davis & Smith, 1992). Often, such conferences advertise themselves as an opportunity for students to make professional-style presentations in a supportive, low-stress atmosphere. According to Carsrud (1975), these meetings have several benefits for students: developing feelings of competence and familiarity with a research area, providing a chance to demonstrate this competence to others, instilling a feeling of excitement about research, providing reinforcement for conducting research, and developing communication skills.

Mount Holyoke College (South Hadley, MA) claims to have both the most well-known (http://www.mtholyoke.edu/acad/psych/orgs.html) and oldest (http://138.110.28.9/offices/student-programs/org_lists/alpha/p.shtml) undergraduate psychology conference in the nation. According to Davis and Smith (1992), the Mount Holyoke conference began in 1948, which would make it older than some of the regional psychology associations for professionals (e.g., Southeastern Psychological Association, founded in 1955; Southwestern Psychological Association, founded in 1954).

Tracking the history of student research conferences in psychology is not an easy task—there are simply not many relevant publications. Starting in 1975, in its second year of publication, *Teaching of Psychology* has published a listing of student psychology conferences. From 1975 to 1991, this list averaged about 11 such conferences per year.

Today, listings for student psychology conferences are more likely to appear on the Web, probably because of the short time lag for a conference posting to appear. The two prime listing sites are the Psi Chi Web site (http://www.psichi.org/conventions/, where you can click on "Student Conventions") and the Society for the Teaching of Psychology's site (http://teachpsych.lemoyne.edu/teachpsych/div/conferences-undergraduate.html). Davis and Smith (1992) alluded to the growth of student research conferences in the 1970s; based on the Web listings, it appears that a second explosion of growth has taken place since that time. A search of the two Web sites in May 2005 turned up listings for 39 different student research conferences held from 2003 onward. Figure 15.1 shows a map of states that had a student research conference during this time interval. Several states had more than one such conference: California (6), Ohio (3), and North Carolina (3) topped that list. The geographical spread of student research conferences seems broader than a few years ago: Davis and Smith noted that such conferences were clustered in the eastern United States. Although the map makes it clear that there are

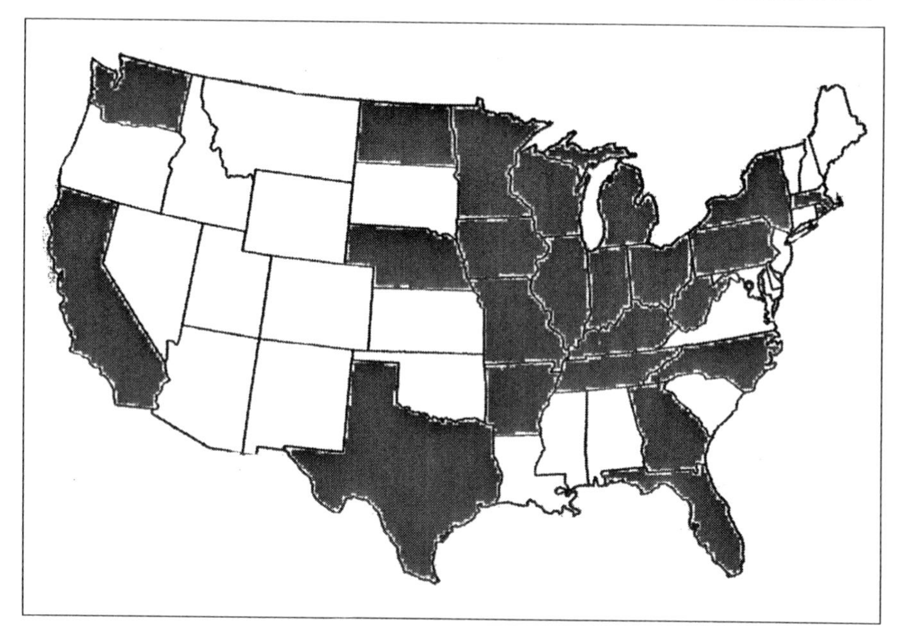

Figure 15.1. State Hasting Psychology Student Research Conferences 2003–2005.

geographical areas that are isolated with respect to student research conferences, many students around the United States have such a conference within driving distance.

Publishing Opportunities for Undergraduate Psychology Students

Graduate schools often stress the importance of research experience for gaining admission to graduate programs (Lawson, 1995; Smith, 1985). That information is consistent with the previous section regarding undergraduate research conferences. Going a step beyond a research presentation to a research publication is even better (Keith-Spiegel, Tabachnick, & Spiegel, 1994; Landrum, Davis, & Landrum, 2000). In fact, Keith-Spiegel et al. (1994), in a survey of 123 graduate faculty members, found that "research experience resulting in a publication credit in a scholarly journal" (p. 80) was the highest ranked criterion for making admissions decisions of 31 items on the survey.

Although undergraduate students applying for graduate school would probably love to have a publication in a top scholarly journal, the likelihood of that type of publication is quite low. As Ferrari, Weyers, and Davis (2002) noted, the process of publication in such a journal is a long, drawn-out process with a low probability of success (i.e., high rejection rates at such journals). Given these obstacles, students are not likely to even begin the publication process, although they may do so in collaboration with a faculty member.

Over the past few years, student psychology journals have appeared in an attempt to provide publication opportunities for undergraduate psychology students. Smith and Davis (2007) provided a list of five journals that are geared toward student research articles. Four of the five have been in existence for more than 10 years, so it is likely that they are here to stay.

There is some disagreement about whether publications in student research journals are helpful for students who are applying for graduate admission. Although Keith-Spiegel (1991) and Powell (2000) gave positive opinions on this question, some data imply that the answer is no. Ferrari et al. (2002) conducted two surveys of college faculty. In the first survey, they listed the five student journals that Smith and Davis (2003) enumerated and assessed faculty familiarity with those journals. Ferrari et al. also asked respondents whether they would advise students to publish in those outlets. The data showed that most faculty were not familiar with the journals and would not advise students to publish in the journals. However, a follow-up survey of 412 American Psychological Society members (Study 2) showed different results. In this survey, Ferrari et al. asked respondents to compare two hypothetical applicants for a PhD program. The students were comparable except that one had published an article in a student psychology journal and the other had not. In another comparison, one hypothetical student had published in a general professional journal and the other had not. In this study, respondents indicated a strong positive reaction toward the publication, even if the publication was in a student journal. It seems that the first survey may have biased respondents against the student journals by first making respondents think about whether they had heard of the journal. Thus, it does appear that undergraduates who have a publication of any sort benefit when applying to graduate school.

How to Maximize Students' Research Experience

Much of the advice that follows may be somewhat idiosyncratic. However, from conversations I have had with colleagues around the country as well as from some evidence I have been able to gather, I know that I am not the only teaching psychologist who holds these beliefs. Christopher and colleagues (chap. 13, this volume) voiced my sentiment well when they wrote about "a specific course requirement that we hold in the highest esteem: the student research project" (p. 223). Indeed, there are many faculty who share the belief that a student research project is one of the best learning experiences students can have. I am one of those faculty. For 26 years, I taught a three-course sequence at a small, private, liberal arts college. Students first took a Statistics course, followed by Experimental Psychology, and Research Methods. In the experimental course, students learned the typical information about conducting psychological research. The culminating project for that course was an experimental research proposal. The following semester, in Research Methods, students actually carried out their proposals. From that experience, I formed some opinions about best practices in teaching those courses that were validated many times over with hundreds of students. Following are several principles that I believe are best practices and that will help students and faculty make the most out of this experience.

1. *Require students to conduct research projects.* In this chapter, I have summarized benefits accruing to students who conduct research projects and subsequently apply to graduate school. Alumni surveys (e.g., Beins, personal communication, November 15, 2005; Lunneborg & Wilson, 1985) have shown that alumni consider courses in statistics and research to have been valuable courses for them, regardless of whether they attended graduate school. Learning to approach problems systematically, looking for background information, gathering data bearing on a question, and drawing conclusions from those data are skills that are useful in a variety of settings and occupations.

2. *Have students conduct original research projects.* Some research-oriented courses have students conduct lab studies in somewhat the same manner that students typically do in a chemistry or physics laboratory. Such canned lab studies simply replicate well-known and reliable findings rather than asking new questions. In my opinion, this type of training is good as a beginning step, but not if it is all the training that students receive. To me, part of the thrill of conducting research is that the outcome is unknown—that the researcher is advancing the state of knowledge in psychology in gathering and analyzing the data. When my students analyzed their data in my presence, the excitement they showed when they got their results from the computer contrasted with the ho-hum attitude that students often show in canned laboratory settings.

3. *Have students work on individual research projects.* I know that many faculty or departments use group projects rather than individual projects. I am a firm believer in students creating individual projects, at least one time in their research careers. I believe that students derive benefits from individual projects that they will not gain from a group project. For example, each student must develop an idea for an original piece of research. As anyone who teaches students how to do research is aware, developing their own research idea is one of the most difficult steps for students—they often believe that they have no ideas or ability to come up with an idea. However, once they have generated an idea, the research project begins to take on a life of its own. Students become more interested in research when they are tackling an idea that they developed. The lack of excitement is one of the biggest problems I see with the canned labs that I mentioned earlier. As faculty, I believe that we want to convey the excitement about

research that we possess; I believe that students working on original, individual research projects are much more likely to develop and show this type of excitement.

4. *Require students to make a written report of their experiment.* I was tempted not to list this step in the research process in the chapter. After all, writing a report after an experiment seems so logical and commonplace; I would like to assume that all faculty do have students engage in this step. However, I believe that it is so vital that students learn the first step in the communication process that I could not leave out this step. Even if you teach an introductory methods course with canned projects, requiring students to compile a lab report is an important step. If you can convince students of the importance of sharing their findings early in their research careers, then the subsequent steps will be easier to manage. Requiring a lab report or experiment write-up begins the process of teaching students how psychologists communicate, how to translate the steps they followed in the research process into written format, and how to use the language and style that psychologists use in their manuscripts and articles. Once a student has used the typical American Psychological Association style and format in writing a report, that student should be better able to extract the important information from a journal article.

5. *Require (or strongly encourage) students to make an oral presentation of their research.* Although writing a research manuscript is a large part of the communication process involving research, it is not the final form of communication. Sharing results with other psychologists is a large part of what researchers do; presenting one's research at a conference is often the first step that we take as professionals. Students can benefit greatly from having that same experience. Thus, I always strongly encouraged my students to attend the statewide student research meeting and to make an oral presentation of their research. I used grades as a strong encouragement; I guaranteed students no lower than a B for their research paper if they actually made a presentation. For many students, making an oral presentation is an intimidating idea; many people are afraid of speaking in public. However, in over 20 years of taking students to the Arkansas Symposium for Psychology Students, I never had a student say afterward that he or she regretted having given that presentation. Instead, they felt a real sense of accomplishment in having done so—something I hope followed them into life in which they might need to speak before a group of people (regardless of graduate school attendance). In addition, they saw other students from around the state give presentations and were able to compare our school to those other schools; typically, they came away feeling quite good about not only themselves, but also about our psychology department and the training it afforded them. During my last few years at my former in-

stitution, the college sponsored a Scholars' Day program. For this program, I urged my students to present their research in a poster format. In this way, my students had the experience of giving both an oral and a poster presentation, and they were able to compare and appreciate the strengths and weaknesses of both approaches. Most students decided that they liked the poster format because of its lower degree of formality and anxiety and because they were able to talk one-on-one with interested observers. Even in a poster format, students gained in self-confidence and a sense of accomplishment. These are valuable lessons for life.

To me, the benefits gained from presentation in either format are important enough to merit taking students to student-oriented conferences. If your region, state, or campus does not have such an outlet, I urge you to work on developing one. Students who make such presentations are likely to benefit in terms of graduate school admissions (Ferrari et al., 2002; Keith-Spiegel, 1991; Powell, 2000). In addition, faculty who have organized such a conference have had good things to say about their experiences (see, e.g., Davis & Smith, 1992; McKenna, 1995).

6. *Encourage some students to submit manuscripts for publication.* Just as making a presentation takes a student a step farther along the professional scholarly research road, submitting a manuscript for publication will give an undergraduate a pretty full glimpse into a research psychologist's existence. Obviously, publishing is not a venture for all undergraduate psychology majors, but there are some undergraduates who will benefit greatly from the process. One obvious group who would benefit is your best majors. These are the students who are definitely graduate-school material. Even among these students, however, are there only some who are ready for the publication process. The choice few are those with strong writing skills and a good grasp of the research process. You will likely have some students who fit these criteria who are not necessarily in the top tier of your students, however. As you well know, the students who are most oriented toward research do not have to be your top students. How, then, do you decide which students should pursue publication?

In my experience, one of the most important criteria I used to differentiate students who might be able to publish was the student's research project and subsequent findings. Often, I got my first inkling of publishing potential when students came to their first or second meeting with me about their future research project. Some students seemed almost to be naturals. Rather than coming in with a standard, tired research idea (e.g., "What happens if students listen to music while they study?"), these students came in with ideas that had theory or theoretical questions behind them. These students were off to a great start. Some of these students turned out to be real go-getters when it came to research. I did not have to push them week after week to start their projects or collect data or score the data—these students were always asking me "What's next?" Finally, some of these students ended up with projects that worked out well; the data they collected proved to be noteworthy in some respect. Perhaps the data shed light on the theoretical question they asked, perhaps the data were simply interesting, or perhaps their interpretation of the data was compelling. Thus, you can see that when I talk about students with publication potential, I am speaking of a subset of a subset of a subset of students. I believe we do our students no favors, we do ourselves no favors, and we do our discipline no favors if we push all or even a majority of our students toward publication. Instead, we should reserve that privilege for the students who truly earn and deserve it. It is important to be realistic with these students; publications are never guaranteed. Although psychology faculty are aware of this fact, it may come as a surprise to the better students who are accustomed to meeting academic challenges successfully. However, even a good student who receives a letter of rejection will benefit from the quest. It is vital for the faculty member to be prepared to serve as a dependable mentor at that point.

CONCLUSION

The psychology curriculum has evolved over the years to include both statistics and research courses. Although students may not like these courses, most faculty and departments appear to believe that these courses are foundational to the discipline. Thus, the belief that conducting a research project is important appears to be a core belief in many departments. However, if students only conduct a research project, they may be getting shortchanged. Adding the additional steps of completing a write-up of the research, presenting the research, and perhaps even publishing the research gives students an in-depth look at the scientific side of psychology and yields many additional benefits to the students. I strongly encourage departments to consider adding these steps to the research process if they have not done so already.

REFERENCES

Baum, C., Benjamin, L. T., Jr., Bernstein, D., Crider, A. B., Halonen, J., Hopkins, J. R., McGovern, T. V., McKeachie, W. J., Nodine, B., Reid, P. T., Suinn, R., & Wade, C. (1993). Principles for quality undergraduate psychology programs. In T. V. McGovern (Ed.), *Handbook for enhancing undergraduate education in psychology* (pp. 17–20). Washington, DC: American Psychological Association.

Brewer, C. L. (1997). Undergraduate education in psychology: Will the mermaids sing? *American Psychologist, 52,* 434–441.

Brewer, C. L., Hopkins, J. R., Kimble, G. A., Matlin, M. W., McCann, L. I., McNeil, O. V., Nodine, B. F., Quinn, V. N., & Saundra. (1993). In T. V. McGovern (Ed.),

Handbook for enhancing undergraduate education in psychology (pp. 161–182). Washington, DC: American Psychological Association.

Buxton, C. E., Cofer, C. N., Gustad, J., MacLeod, R. B., McKeachie, W. J., & Wolfle, D. (1952). *Improving undergraduate instruction in psychology*. New York: Macmillan.

Carsrud, A. L. (1975). Undergraduate psychology conferences: Is good research nested under Ph.D.s? *Teaching of Psychology, 2,* 112–114.

Daniel, R. S., Dunham, P. J., & Morris, C. J., Jr. (1965). Undergraduate courses in psychology; 14 years later. *The Psychological Record, 15,* 25–31.

Davis, S. F., & Smith, R. A. (1992). Regional conferences for teachers and students of psychology. In A. E. Puente, J. R. Matthews, & C. L. Brewer (Eds.), *Teaching psychology in America: A history* (pp. 311–323). Washington, DC: American Psychological Association.

Ferrari, J. R., Weyers, S., & Davis, S. F. (2002). Publish that paper—But where? Faculty knowledge and perceptions of undergraduate publications. *College Student Journal, 36,* 335–343.

Henry, E. R. (1938). A survey of courses in psychology offered by undergraduate colleges of liberal arts. *Psychological Bulletin, 35,* 430–435.

Keith-Spiegel, P. (1991). *The complete guide to graduate school admission: Psychology and related fields.* Hillsdale, NJ: Lawrence Erlbaum Associates.

Keith-Spiegel, P., Tabachnick, B. G., & Spiegel, G. B. (1994). When demand exceeds supply: Second-order criteria used by graduate school selections committees. *Teaching of Psychology, 21,* 79–81.

Kierniesky, N. C. (1984). Undergraduate research in small psychology departments. *Teaching of Psychology, 11,* 15–18.

Kierniesky, N. C. (2005). Undergraduate research in small psychology departments: Two decades later. *Teaching of Psychology, 32,* 84–90.

Kulik, J. A. (1973). *Undergraduate education in psychology*. Washington, DC: American Psychological Association.

Landrum, R. E., Davis, S. F., & Landrum, T. (2000). *The psychology major: Career options and strategies for success.* Upper Saddle River, NJ: Prentice-Hall.

Lawson, T. J. (1995). Gaining admission into graduate programs in psychology: An update. *Teaching of Psychology, 22,* 225–227.

Lunneborg, P. W., & Wilson, V. M. (1985). Would you major in psychology again? *Teaching of Psychology, 12,* 17–20.

Lux, D. F., & Daniel, R. S. (1978). Which courses are most frequently listed by psychology departments? *Teaching of Psychology, 5,* 13–16.

McGovern, T. V., Furumoto, L., Halpern, D. F., Kimble, G. A., & McKeachie, W. J. (1991). Liberal education, study in depth, and the arts and sciences major—Psychology. *American Psychologist, 46,* 598–605.

McGovern, T. V., & Reich, J. N. (1996). A comment on the *Quality Principles*. *American Psychologist, 51,* 252–255.

McKeachie, W. J., & Milholland, J. E. (1961). *Undergraduate curricula in psychology*. Fair Lawn, NJ: Scott, Foresman.

McKenna, R. J. (1995). The undergraduate researcher's handbook: Creative experimentation in social psychology. Needham Heights, MA: Allyn & Bacon.

Messer, W. S., Griggs, R. A., & Jackson, S. L. (1999). A national survey of undergraduate psychology degree options and major requirements. *Teaching of Psychology, 26,* 164–171.

Perlman, B., & McCann, L. I. (1999a). The most frequently listed courses in the undergraduate psychology curriculum. *Teaching of Psychology, 26,* 177–183.

Perlman, B., & McCann, L. I. (1999b). The structure of the undergraduate psychology curriculum. *Teaching of Psychology, 26,* 171–176.

Perlman, B., & McCann, L. I. (2005). Undergraduate research experiences in psychology: A national study of courses and curricula. *Teaching of Psychology, 32,* 5–14.

Powell, J. L. (2000). Creative outlets for student research or what do I do now that my study is completed? *Eye on Psi Chi, 4*(2), 28–29.

Sanford, F. H., & Fleishman, E. A. (1950). A survey of undergraduate psychology courses in American colleges and universities. *American Psychologist, 5,* 33–37.

Scheirer, C. J., & Rogers, A. M. (1985). *The undergraduate psychology curriculum: 1984.* Washington, DC: American Psychological Association.

Smith, R. A. (1985). Advising beginning psychology majors for graduate school. *Teaching of Psychology, 12,* 194–198.

Smith, R. A., & Davis, S. F. (2007). *The psychologist as detective*: An introduction to conducting research in psychology (4th ed.). Upper Saddle River, NJ: Prentice Hall.

Task Force on Undergraduate Psychology Major Competencies. (2002). *Undergraduate psychology major learning goals and outcomes: A report.* Retrieved May 30, 2005 from http://www.apa.org/ed/pcue/taskforcereport.pdf

Part VI

Special Topics: Diversity Issues and Writing

Chapter 16

Understanding the Mosaic of Humanity Through Research Methodology: Infusing Diversity Into Research Methods Courses

Linda M. Woolf and Michael R. Hulsizer
Webster University

Scanning the tapestry of human existence, one factor becomes immediately clear—humanity is marked by both incredible diversity and similarity. Unfortunately, much of the psychology curriculum has historically neglected or offered only a cursory glance at issues of diversity (Guthrie, 1998). However, in the early to mid-1990s, the American Psychological Association (APA) began developing strategies for incorporating diversity into the curriculum. According to Kowalski (2000), prior to the APA's involvement, the most prevalent approach toward infusing multiculturalism into the curriculum was the occasional addition of courses that focused exclusively on specific minority groups (e.g., women) or shared minority experiences (e.g., racism, prejudice and discrimination).

In 1998, APA's Board of Educational Affair's Task Force on Diversity Issues at the Precollege and Undergraduate Levels of Education in Psychology authored a series of articles in the *APA Monitor* http://www.apa.org/ed/divhscollege.html, which suggested that a more effective approach to making psychology more inclusive would be to incorporate diversity issues within existing courses in psychology. Consequently, psychologists have increasingly integrated material related to gender, race, and multicultural concerns into the psychology curriculum. The American Psychological Association's creation of the *Guidelines on Multicultural Education, Training, Research, Practice, and Organizational Change for Psychologists* (2003) as well as the formation of the 2005 Education Leadership Conference devoted to diversity issues further highlighted this trend.

Quina and Kulberg (2003) highlighted five goals and objectives that accompany the integration of sociocultural context and multicultural awareness into an experimental psychology course. These objectives include (a) teaching appropriate research design selection, (b) strategies for reducing bias in relation to both research design and the interpretation of results, (c) techniques to improve external and internal validity, (d) ethics related to research design and interpretation, and (e) how to be informed consumers of research.

In addition to these concrete goals, there are other related benefits associated with integrating diversity concerns into a research methods course. First, the integration of diversity issues into a discussion of research methods can facilitate further development of critical thinking skills. For example, research related to diversity has demonstrated that some psychological concepts and theories do not have universal applicability. Further, the traditional description and recommended treatments for posttraumatic stress disorder do not have cross-cultural applicability (Atlani & Rousseau, 2000; Bracken, 1998). Such knowledge can challenge students' notions of "truth" and can highlight the need for students to approach critically the study of research methods. Second, there is growing awareness that the use of homogeneous samples (e.g., introductory psychology students), which has a long tradition in psychology, has painted a somewhat distorted picture of psychological phenomena (Graham, 1992; Sears, 1986). Thus, there is a need to train the next generation of psychologists to avoid such pitfalls. Third, only through research related to human variability can students begin to examine the commonalities and differences in human behavior around the globe. Finally, the study of diversity within a research methods course opens the door to the inclusion of diversity topics into their pool of potential areas of research.

The remainder of this chapter is divided into common categories of *diversity* (race/ethnicity, gender, etc.). After a brief introduction, each section highlights the means by which a research methods course can benefit from the infusion of specific diversity areas. For example, international research can be discussed in relation to problems associated with something seemingly as simple as developing demographic questions; age-related research can be introduced during a discussion of quasiexperimental design concerns; and research involving people with disabilities can be discussed when examining the distinction between performance and ability. However, even though each particular diversity category contains a discussion of specific research issues, these issues merit consideration in many domains. Thus, issues we raise here concerning quasiexperimental designs are not unique to age-related research. Rather, we discuss these issues primarily within that context by way of example.

Although we focus on six areas of diversity (i.e., race/ethnicity, gender, sexual orientation, age, disability, cross-cultural/international), it is important to remember that these areas are neither mutually exclusive nor homogeneous categories. It is also important to note that we do not discuss all issues related to diversity concerns and research. Rather, representative research topics and examples demonstrate the complexity involved in conducting research responsibly— particularly research that involves diverse populations. The accompanying CD-ROM includes suggestions for class exercises designed to demonstrate diversity-related research concerns.

RACE AND ETHNICITY: THE LESSONS OF EXCLUSION AND GROUP DEFINITION

The United States Census Bureau (2004) projected that the number of non-White Americans would increase from approximately 20% of the population in 2000 to almost a third of the population by 2050. In addition, during the same period, the U.S. Census Bureau reported that the number of Hispanics (of any ethnicity) is expected to double, eventually comprising almost a quarter of the population by 2050. According to data collected in October 2003 by the U.S. Census Bureau (Shin, 2005), approximately 30% of college students were African American, Asian, or Hispanic. Indeed, these population trends mirror the diverse nature of U.S. classrooms. When one begins to look internationally, the tapestry only becomes more complex.

Although the nature of society is changing, it is still the case that ethnic minority groups are rarely the focus of empirical research (Sue, 1999). For example, several researchers have noted the paucity of published research that focus on racial or ethnic minorities within psychology journals (e.g., Bernal, Trimble, Burlew, & Leong, 2003; Carter, Akinsulure-Smith, Smailes, & Clauss, 1998; Graham, 1992). Additionally, some authors have examined the degree to which researchers are simply willing to report the racial and ethnic background of study participants (e.g., Buboltz, Miller, & Williams, 1999; Munley et al., 2002). In a recent study, Delgado-Romero, Galván, Maschino, and Rowland (2005) found that only 57% of examined counseling articles published between 1990 and 1999 reported race and ethnicity. The authors concluded that, as compared to the U.S. population, White and Asian Americans were overrepresented in counseling research, whereas African Americans, Hispanics, and American Indians were underrepresented. It would seem that Betancourt and López's (1993) assertion that that "the study of culture and related variables occupies at best a secondary place in American (mainstream) psychology" (p. 629) is as true today as it was over a decade ago.

In order to enable investigators to conduct research using diverse populations, the five ethnic minority associations that compose the Council of National Psychological Associations for the Advancement of Ethnic Minority Interests (CNPAAEMI) worked collaboratively to create the *Guidelines for Research in Ethnic Minority Communities* (2000). Consequently, the *Guidelines*, which provide a wealth of information concerning research with African American, Asian American, Hispanic, and American Indian populations, is an extremely useful addition to any research methods course. However, it is important to keep in mind that although we make many of the suggestions that follow within the context of a specific ethnic minority group, these suggestions may be applicable to other group experiences. Indeed, researchers need to take care when using race as a demographic category, lest the variable become a proxy for biological explanations of behavior (Wang & Sue, 2005). Nonetheless, the race variable is a focal social and cultural construct that, if used judiciously, can provide a wealth of information. In fact, one common recommendation proposed by the *Guidelines for Research in Ethnic Minority Communities* invites researchers to learn more about the rich diversity among individuals defined as belonging to a particular race and avoid treating racial categories as homogenous groups.

The *Guidelines for Research in Ethnic Minority Communities* (CNPAAEMI, 2000) address several issues concerning research with persons of African descent that a research methods course can address. First, students need to be aware of the impact of racism in America when researching topics in which Blacks have been historically disenfranchised, such as employment and education. Second, instructors may also want to define the "one drop rule"—the tendency to label individuals within the United States as Black if their ancestry included anyone considered to be of African-American descent. A critique of this rule can serve as a nice jumping-off point for a discussion of the perils of research designs that compare racial groups (see Wang & Sue, 2005). In addition, a discussion of the Human Genome Project may be a useful way to reinforce the point that there is more diversity within racial groups than between them (Rosenberg et al., 2002; Zuckerman, 1990). Students need to recognize that the concept of *race* is in large measure a social construction. Third, when comparing persons of African descent and other racial groups (typically a White control group), it is important to impress on students that it is not appropriate to assume that deviations away from the control group reflect undesirable characteristics. Finally, a lecture on research ethics would be incomplete without a discussion of the infamous Tuskegee syphilis experiment (Jones, 1993).

In discussing the issues surrounding the use of Hispanics in research, the *Guidelines for Research in Ethnic Minority Communities* (CNPAAEMI, 2000) state that, above all else, researchers need to consider the breadth of diversity within the Hispanic population. Specifically, this broadly defined group varies with respect to language, race/ethnicity, acculturation, beliefs, socioeconomic status, and educational background. These same concerns are also at the core of the discussion of research guidelines associated with American Indians. Depending on the tribe, research conducted with American Indians will often involve the use of small sample sizes. Although small sample sizes limit the ability of the researcher to find statistical significance, more homogenous Pan-Indian designs distort the cultural uniqueness of each tribe.

The *Guidelines* (CNPAAEMI, 2000) raise several important assumptions and methodological concerns inherent in conducting research with Asian American/Pacific Islanders. First, it is important to understand the diversity of cultures that fall within the category of *Asian American/Pacific Islander*—particularly given existing research takes place primarily with Chinese and Japanese American participants. Second, the inclusion of an Asian American/Pacific Islander investigator or collaborator increases the validity of research with Asian Americans/Pacific Islanders. For example, this inclusion can better enable the research team to ask the right questions. Third, assessment instruments and measures need to be validated using each cultural population (additional information regarding this process appears in the cross-cultural/international section). Finally, researchers need to take into account historical context when drawing conclusions using Asian-Americans/Pacific Islanders. These points may prove to be useful for stimulating class discussion and enhancing awareness of the issues surrounding research with culturally diverse populations.

GENDER: THE LESSONS OF EXPERIMENTAL BIAS

Since 1979, women have consistently been the majority presence in college classrooms (Shin, 2005). Indeed, the American Psychological Association Research Office (Pate, 2001) reported that women comprised 71% of first year U.S. psychology graduate students (75% of Canadian psychology graduate students) in 1999–2000. However, the Research Office report also stated that during the same period, only 37% of full-time faculty in public and private graduate psychology departments within the United States and Canada were women. Although the report stated that the discrepancy between male and female full-time graduate psychology faculty members has decreased since 1984 (when men composed 78% of full-time faculty positions), the fact remains that women are not well represented in the training of future psychologists.

Not surprisingly, critics see the disparity between male and female full-time psychology graduate faculty as but one instance in a long and continually evolving list of examples demonstrating the dominant androcentric bias in mainstream psychology (Sherif, 1979; Weisstein, 1971; Yoder & Kahn, 1993). One area that has received considerable attention has been the impact this gender bias has had on research methodology. For example, Eichler (1991) described several "sexist problems" in research. Although the issues Eichler cited were not specific to psychology, they can be discussion points for use in a psychology research methods class.

According to Eichler (1991), one impact of cultural sexism on research is *androcentricity*—the tendency for researchers to view the world from a male perspective. The research process perpetuates androcentrism in many ways. In an influential article, Denmark, Russo, Frieze, and Sechzer (1988) discussed several examples of sexism in research and suggested means by which researchers can avoid such problems. For example, the authors discussed the tendency for researchers to examine concepts such as *leadership style* in a fashion that emphasizes male stereotypes (e.g., dominance, aggression) versus a range of leadership styles, including styles that are more egalitarian. Another example cited by Denmark and colleagues involves the propensity for some researchers to assume that topics relevant to White males are inherently more important than issues related to other groups.

Eichler (1991) cited *overgeneralization*, the propensity to extend concepts, theories, and research developed using one sex as indicative of the behavior of all individuals, as another common problem in research. Examples of overgeneralization range from the use of sexist language and biased research methodology to the tendency of researchers to generalize the results of a particular line of inquiry to all individuals when, in fact, the initial sample was only composed of men. In addition to overgeneralizing results, some researchers also employ the "female deficit model" when drawing conclusions (Hyde, 1996). That is, the tendency to frame the results in such a fashion that women's behavior is seen as deficient or nonnormative.

Gender insensitivity, the propensity for researchers to ignore gender as an important variable of interest, is the third issue present in sexist methodology (Eichler, 1991). Indeed, Denmark and colleagues (1988) called for researchers to cite the gender of research participants, experimenters, and anyone else that may be active participants in the study (e.g., confederates). It appears that Denmark and colleagues' call for action has had an impact. According to Munley et al. (2002), approximately 89% of research articles the authors examined in 1999 reported the gender of participants.

Instructors may find it useful to temper students' enthusiasm toward interpreting gender sex differences as indicative of a continuum with masculine and feminine attributes as bipolar opposites (McHugh, Koeske, & Frieze, 1986). Indeed, McHugh and Cosgrove (2002) suggested that by attributing differences in men and women to gender differences, "we risk essentializing gender, promoting a view of women as a homogenous group, and reinforcing the very mechanisms of oppression against which we are fighting" (p. 13). Moreover, it is important to remember that gender roles, masculine or feminine, are neither static nor the sole purview of either men or women, respectively. Additionally, the use of only *male/men* or *female/women* as categorical terms is exclusionary of those who are members of the transsexual, transgender, or intersex communities.

Meta-analysis has become an increasingly popular tool for psychologists. A search of PyscInfo© revealed dozens of meta-analyses examining gender differences across a variety of issues. Although meta-analyses have almost certainly added to psychologist's knowledge of humanity, researchers need to exercise caution when drawing conclusions from such studies. For example, Halpern (1995) expressed concern that researchers are ignoring context when using benchmarks to determine whether a gender-related effect size is meaningful or important. The tendency to ignore context is compounded by the fact that studies investigating gender differences often do not use very representative samples. Halpern cautioned that "as long as the research literature is based on a narrow selection of participants and assessment procedures, the results of any analysis will be biased" (p. 83).

Hyde (1996) offered four characteristics of "gender-fair research." First, never conduct research using just one gender. Second, researchers should assess all assumptions, hypotheses, and theoretical models to assure they are gender-fair. Third, researchers should be aware that the context within which they collect data could have an impact on the results. Consequently, it is important to use male and female experimenters and confederates when at all possible. Finally, researchers should exercise caution when interpreting research results, particularly if the findings suggest gender differences.

SEXUAL ORIENTATION: THE LESSONS OF SAMPLING

Attitudes concerning the subject of sexual orientation have undergone significant change in the past 20 years. According to recent Gallup Polls (Saad, 2005), Americans have become increasingly more tolerant of homosexuality. In 1982, only 34% of participants agreed that homosexuality was an acceptable alternative lifestyle. This number had risen to 54% when participants responded to the same question in 2004. The rise in acceptance of homosexuality is most pronounced among college-age respondents. According to a 2002 Gallup Poll, 62% of young adults (ages 18 to 29) considered homosexuality acceptable, and 65% believed homosexual relations should be legal (Carlson, 2002).

Despite these changing attitudes, research inclusive of or concerning sexual orientation remains a small fraction of the literature and is contained largely in specialized journals or special issues of journals (Bowman, 2003). One of the major factors limiting an individual's ability to conduct research on this topic is the difficulty of sampling from a hidden population. This difficulty challenges researchers studying any number of populations (e.g., the mentally ill) who are not immediately identifiable based on some external characteristic. The challenge is only magnified when the group involved belongs to a class of individuals that is socially stigmatized because of that classification. Many individuals may be reticent to self-identify as lesbian, gay, or bisexual (LGB) due to fears concerning confidentiality or internalized homophobia. Thus, researchers are often limited to a very select group of the population who self-identify as LGB—typically White, formally educated, urban, and upper middle class.

In addition to the hidden population challenge, the terms *sexual orientation* (e.g., gay, lesbian, bisexual) and *gender identity* (e.g., transgendered, intersex) are often confused as both may involve groups commonly identified as sexual minorities. It is not uncommon to see the following acronym when reading materials related to or written by those within sexual minority communities in the United States—GLBTQI (gay, lesbian, bisexual, transgendered or transsexual, queer or questioning, intersex). This juxtaposition of identity and orientation may be socially beneficial for individuals experiencing stigma in the United States, but can create the basis for problematic research hypotheses and, certainly, sampling challenges.

Additionally, there is no uniform consensus over the terms used for sexual orientation group identification, or even which terms society deems acceptable or appropriate. For example, individuals often prefer the terms *gay* or *lesbian*, given the fact that homosexuality has historically been associated with disease and mental illness. Today, others are replacing the terms gay and lesbian with *queer*, as they believe this expression is more inclusive of bisexuals and transgendered individuals (Grace, Hill, Johnson, & Lewis, 2004). Even more challenging, from a research perspective, is that some individuals who have had same-sex sexual relationships may still self-identify as heterosexual (Savin-Williams, 2001). This ambiguity and interplay between all of these issues makes simple, nonfluid sexual orientation, self-identification more ambiguous and a challenge when attempting to classify research participants based on sexual orientation.

Finally, it is important to be aware of heterosexist bias in research that may deter individuals from participating in research or skewing research results. One simple way to bring this point home with students is to introduce them to Rochlin's (1995) Heterosexual Questionnaire, which includes items such as "What do you think caused your heterosexuality?" In response to the challenges associated with obtaining a representative sample and avoiding heterosexist bias, researchers are currently exploring new methodologies such as the use of the Internet for research purposes (Koch & Emrey, 2001) and the use of alternative and qualitative methodologies related to the study of sexual orientation (Morrow, 2003).

AGE: THE LESSONS OF METHODOLOGICAL DESIGN

Over the past century, the population of the United States has undergone a dramatic demographic shift with respect to aging. The traditional age pyramid, with a large number of children forming the base and the oldest citizens representing a small point at the top, has become increasingly rectangular. Currently, 12.4% the population of the United States is aged 65 or older, and researchers estimate that this figure will increase to approximately 20% by the year 2030 (Administration on Aging, 2004). The oldest segment of the population, those 85 and older, is anticipated to more than double during that same period. According to a report on global aging (Kinsella & Phillips, 2005), this graying of the population is not confined to the United States or even to developed countries around the globe. In fact, changes in fertility rates, reductions in infant mortality, increased life expectancy resulting from better access to clean water, availability of health care, and other factors has resulted in a global aging phenomenon. Kinsella and Phillips reported that there are currently at least 461 million people aged 65 or older around the globe with this number anticipated to increase to over 970 million over the next 25 years. In addition, 60% of the global aging population lives in less developed countries. By the year 2030, Kinsella and Phillips anticipate this figure will rise to 71%.

Research concerning older adults has burgeoned; when one evaluates such research, there are several important issues to consider. First, *aging* is not a unidimensional concept. Rather, aging can occur along several dimensions (e.g. chronological, biological, psychosocial) and should be viewed as multidirectional, multidimensional, plastic, and contextual (Lerner, 1995). Unfortunately, most research relies on chronological age as the primary criterion in evaluating research participants' age.

A second concern relative to aging research is the tendency toward studying older adults as a homogeneous population. In fact, the older population is marked by incredible diversity. Even on the most basic level, there are at least two age-cohorts commonly identified: young-old and old-old/frail elderly. Baltes (1997) described this later age category as a "major new frontier for future research and theory in life span development as well as for efforts in human development policy" (p. 375). Clearly, there is a difference between studying a healthy, recently retired group of individuals in their 60s and a group of individuals in their 90s who may be struggling with chronic health concerns. Moreover, as individuals move through the life course, their life experiences, choices, and the multiple dimensions within which they age all foster increasing individuality and variability. Conversely, biology, shared culture, and history can foster consistency with respect to individual life cycles within a community. Thus, one cannot accurately speak of an older adult cohort but rather must think in terms of older adult cohorts.

All of these factors can have an impact on the results of research studies examining aging. The impact is most evident with two commonly used developmental designs: the cross-sectional and the longitudinal designs. The cross-sectional method consists of two or more age cohorts tested at one time to examine differences across ages. Unfortunately, the use of distinct samples in this type of quasiexperimental design typically results in low internal validity. The samples may be different on any number of variables other than the one under investigation. In the cross-sectional study, age differences may be confounded with generational or differences resulting from nonshared cultural and historical cohort experiences. Therefore, differences between age groups can be described but not explained. For example, early cross-sectional studies suggested progressive declines in intelligence following late adolescence and early young adulthood. This gloomy picture is not supported by research when taking cohort effects (e.g., differences in educational levels) into account (Schaie, 2000).

The longitudinal method consists of one group of individuals within an age cohort tested at least twice over time. Similar to a time series design, it suffers from many threats to internal validity, with history being the most serious threat. In the longitudinal method, age differences or differences in maturation are confounded by historical events. What occurs in the environment represents an experimental treatment; these events may change people in unexpected ways. For example, the attacks of September 11, 2001, most likely have compromised the results of any current longitudinal research study by introducing a *confound* (e.g., historical events and maturation). Consequently, the results again are largely descriptive.

There are several additional threats to internal validity with the longitudinal method. First, the longitudinal method rarely meets the criteria of selective sampling. For example, individuals who volunteer to participate in a longitudinal study are usually of higher intelligence and socioeconomic status (Baltes, 1968). Second, longitudinal studies suffer from selective drop-out/experimental mortality. Testing effects are also a problem with the longitudinal method. This problem is particularly evident in studies where participants have been retested many times. For example, the Berkeley Growth Study tested the majority of participants approximately 38 times over a period of 18 years (Bayley, cited in Baltes, 1968). Despite these challenges, Schaie (2000) proposed that analyses based on multidisciplinary, multivariate longitudinal studies have the potential to lead to a greater understanding of adult development.

Schaie (1986) also proposed a general developmental model of sequential designs for the analysis of age-related changes. In addition to being a descriptive model, it is a model of theory building. The purpose of these sequential designs is to separate out the variance that is accounted for by normative history (time of measurement), cohort effects, and age. For example, the cross-sequential method allows researchers to examine all cohorts, at all times of measurement, over time. In many ways, the cross-sequential method represents a combined cross-sectional and longitudinal approach. It assumes that time of measurement, cohort, and age are additive and thus the amount of variance due to each can be divided out. However, if interactions occur, it becomes much more difficult to interpret these data.

Finally, sampling issues, reliability and validity of measures, questionnaire wording, experimental bias, and other traditional methodological concerns are important issues to examine when planning or evaluating research relative to aging. For example, research conducted with noncommunity dwelling older adults is unlikely to be a valid representation of the majority older adult population, and measures found to be reliable and valid with a younger adult population may not be so for an older population. Therefore, researchers need to be aware of methodological concerns related to age to avoid the introduction of ageist bias into a research study.

DISABILITY: THE LESSONS OF CONSTRUCT VALIDITY AND INFORMED CONSENT

Researchers estimate that there are almost 77.5 million community dwelling people with disabilities living in the U.S. according to the 2003 American Community Survey (U.S. Census Bureau, 2003). This figure represents approximately 26% of the population. Types of disability are broadly defined to include *sensory, physical, mental, self-care, home-bound,* and *employment* categories. Disabilities can be visible or hidden (e.g., heart defect), and individuals may or may not define themselves as *disabled* depending on the degree of limitation the disability has on their lives. Thus, as with other categories of diversity discussed in this chapter, there is extensive variability within disabled populations.

Although a large portion of the U.S. population lives with some form of disability, research that is inclusive of individuals with varying levels of disability is much less common. There are several reasons for this exclusion as well as potential for bias in the research.

First, measures of assessment used in research, particularly those related to personality assessment and psychopathology, have rarely been standardized for use with a disabled population (Elliott & Umlauf, 1995; Pullin, 2002). Individuals experiencing disability that affects faculties such as motor performance or sensory acuity may face the same tasks and norms as those individuals not experiencing any form of limitation. For example, an individual with a complex of physical disabilities or chronic illnesses might respond to the Hypochondriasis Scale of the Minnesota Multiphasic Personality Inventory (MMPI-2; Butcher et al., 2001) with seeming obsession over body functions. In addition, an individual with arthritic hands may take longer to complete a timed test of intellectual ability. This lack of standardization and potential lack of test accommodation raise the issue of whether individuals are being evaluated more on performance than on ability.

This performance versus ability threat to construct validity also applies to research situations without the use of standardized tests. Researchers rarely design studies to accommodate participants with disabilities. Unfortunately, tasks designed to measure cognitive, personality, or other psychological abilities may instead be measuring primarily noncognitive or nonpsychological variables in research inclusive of participants with disabilities. For example, research on memory using computers to present stimuli may be more challenging for some older adults due to an increased incidence in susceptibility to glare from cataracts or due to presbyopia. It is also possible that some older adults may fatigue more easily or that other health concerns may inhibit performance. Thus, differences found between younger and older participants may reflect visual or health differences as opposed to memory differences. Unless the researcher takes into account the impact of disability, the resulting research conclusions may be biased and inaccurate.

Whereas many researchers fail to accommodate disabled research participants, other researchers simply exclude individuals with disabilities from research participation based on the assumption that a disability has broad effects on cognition, personality, affect, or ability. Unfortunately, this assumption is grounded largely in stereotype rather than reality (Dunn, 2000). According to Dunn, "Insiders (people with disabilities) know what disability is like, whereas outsiders (people without disabilities) make assumptions, including erroneous ones, about it" (p. 574). Most commonly, nondisabled individuals overestimate the effect of a disability on an individual's capacities and well-being when, in fact, situational and interpersonal factors have a far greater impact. Researchers explicitly studying participants with known disabilities need to consider these factors as well as other possible alternative explanations rather than simply attributing the results to the disability. For example, researchers comparing math scores among disabled and nondisabled populations need take into account the role of stereotype threat (Steele, Spencer, & Aronson, 2002) in any performance-based deficits among disabled populations.

Another factor leading to exclusion of individuals with developmental or cognitive disabilities from research is the challenge of informed consent (Dresser, 2001). Informed consent rests on the premise that research participants can fully understand the procedures, risks, and benefits associated with a particular study and agree to participate. Unfortunately, individuals with significant cognitive impairment due to developmental disabilities, injury, or dementia may not fully understand the provided material to give informed consent. Researchers must take special care to protect participants who are unable to provide consent on their own, including situations where a guardian provides consent but the participant is exposed to experimental testing without their knowledge. Due to the potential risks involved, however, many researchers simply avoid research with disabled populations (Yan & Munir, 2004).

CROSS-CULTURAL/INTERNATIONAL: THE LESSONS OF OMISSION AND CULTURAL EQUIVALENCY

Psychology textbooks and journals are filled with psychological concepts, theories, and research findings that, at the surface, appear to be applicable to all humans. Researchers often refer to these universal concepts as *etics*. On the other hand, investigators refer to concepts thought to be specific to a particular culture as *emics*. Although universal and culturally specific findings are present in psychology, Matsumoto

(1994) suggested that "most cross-cultural psychologists would agree that there are just as many, if not more, emics as there are etics" (p. 5). This statement is ironic, given that the majority of research studies described in introductory psychology textbooks are presented without reference to inherent cultural limitations in the research design (Woolf, Hulsizer, & McCarthy, 2002). Indeed, Marsella (2001) suggested that many psychologists are unwilling to "accept a very basic 'truth'—that Western psychology is rooted in an ideology of individualism, rationality, and empiricism that has little resonance in many of the more than 5000 cultures found in today's world" (p. 7).

In fact, it is not surprising that even the most astute student, after reading an introductory psychology textbook, would assume that virtually all research is conducted within the United States. For example, Woolf and colleagues (2002) conducted an analysis of international content in over two dozen popular introductory psychology textbooks. The results revealed that the average percent of international content overall per textbook was only 2.07 (SD = 1.02). Fifteen commonly used lifespan development (M = 3.61; SD = 1.33) and 12 popular social psychology (M = 5.33; SD = 3.85) textbooks yielded similar percentages. We can compare these data to the research results of Adair, Coelho, and Luna (2002), which revealed that 45% of articles indexed in PSYCInfo© included authors from outside the United States. Thus, it is clear that a significant portion of the psychology research literature is missing from the curriculum.

The creation of separate stand-alone classes on cross-cultural/international psychology is one means to increase student's exposure to cross-cultural and international research. However, stand-alone courses do little to infuse the material into mainstream classes (Kowalski, 2000) and may only serve to ghettoize the topic (Quina & Bronstein, 2003). Fortunately, infusing cross-cultural and international research methods into a traditional methods class is relatively simple. According to Matsumoto (2003), "the same problems and solutions that are typically used to describe issues concerning experimental methodology in general can and should be applied to most, if not all, cross-cultural comparisons" (p. 193). Matsumoto went on to state that there "are only a few issues specific to the conduct of cross-cultural research that set it apart from general experimentation" (p. 193).

Experimenters constantly need to monitor the degree to which the methodology is equivalent throughout the cross-cultural comparison. Matsumoto (2003) suggested that the perfect cross-cultural study is an ideal that researchers should strive to meet. Unfortunately, there is no easy way to assess whether researchers achieve that ideal. Brislin (2000) asserted that cross-cultural/international investigators needed to be aware of three sources of nonequivalence—translation, conceptual, and metric. Translational equivalence is necessary when conducting research on one population using experimental measures developed and standardized with another population. So, how should the researcher go about translating a scale or experimental measure? According to the *Guidelines for Research in Ethnic Minority Communities* (CNPAAEMI, 2000), the proper translation of experimental measures necessitates the use of the back translation method. First, a bilingual translator converts the measure to the target language. Next, a second translator converts the translated measure back to the original language. Finally, a researcher compares the original version to the back-translated version to examine any existing differences.

Conceptual equivalence is the degree to which theoretical concepts or constructs are the same between two cultures. Researchers can introduce conceptual nonequivalence in a variety ways, even something as seemingly innocuous as demographics. It is often useful to control for certain demographic variables, such as age, sex, and religious orientation, when conducting cross-cultural research. However, according to Matsumoto (2003), this technique becomes more of a challenge when the issue of culture becomes intertwined with demographics. For example, a demographic question such as "age" would seem relatively culture-free. Yet, for the Ju/'hoansi, also known as the !Kung, this would be not be a useful question as the Ju/'hoansi use a culturally specific age categorization system as opposed to thinking of age in chronological years (Hames & Draper, 2004). Finally, researchers need to be concerned with metric equivalence—the ability to compare the specific scores on a scale of interest across cultures. For example, would a researcher interpret a score of 25 on the Beck Depression Inventory (Beck, Ward, Mendelson, Mock, & Erbaugh, 1961) the same way across cultures?

Not only are these issues of concern when conducting research between cultures, but also within a culture—particularly one as diverse as the United States. For example, according to data collected in October 2003 by the U.S. Census Bureau, almost 22% of college students had at least one "foreign" born parent and 10% of those students were not born in the United States themselves (Shin, 2005). It seems that international research does not have to occur only after crossing large bodies of water.

SPECTRUM OF DIVERSITY

The categorizations discussed previously do not reflect the full spectrum or fluidity of diversity as is present across the vast plains of humanity. Although we have presented six categories commonly discussed in relation to diversity, this discussion represents principally a starting point for discussion and exploration. Other areas of diversity that have a significant impact on individuals' identities, their place in community, and the methods of research include socioeconomic status, language, educational level, religion, marital status, social class/caste, computer literacy, and physical appearance. This listing, of course, is also not definitive, and all of these sources of variability interact. Indeed, identity is not bounded by specific group categorizations but rather is influenced by the currents of culture.

PUSHING THE PARADIGM

In a provocative article, Chang and Sue (2005) suggested that the current scientific paradigm, with its focus on experimental designs, has introduced a bias in psychology. Specifically, mainstream journals systematically exclude multicultural research, thus devaluing the quality of existing multicultural research and impeding the growth of future research in this area. Ideally, experimental research should be high in both *internal validity* (the extent to which causal conclusions can be drawn from empirical results) and *external validity* (the ability to generalize research results to different populations). Unfortunately, it is very difficult to achieve both in experimental designs. Chang and Sue (2005) argued that the importance psychologists have placed on determining causality has led researchers to value internal validity over external validity.

Many research texts discuss the inherent tension between internal and external validity. However, few methodology texts discuss in any detail the consequences that can result from establishing high internal validity at the expense of external validity—particularly as they relate to multicultural research. According to Chang and Sue (2005), the emphasis the discipline places on causality (i.e., high internal validity) facilitates (a) overuse of college students as research participants; (b) the overwillingness to assume research conducted on one population (e.g., White, middle-class, U.S. citizens) can be generalized to other groups and situations; (c) disregard for research that can be seen as seeking to explore cross-cultural differences versus explaining such differences; (d) the tendency of journal reviewers to insist that a White control group is needed when conducting research on ethnic minority groups; and (e) the formation of aggregate non-White populations to obtain a large enough sample size. The latter issue is especially problematic due to the implied assumption of homogeneity among the non-White population. Additionally, although the exclusion of diverse participants may enhance internal or statistical conclusion validity, the resultant homogeneous sample does not reflect the diversity of human experience. Therefore, it is important that psychologists and their students work to be more inclusive of diverse populations in all of their research.

FINAL THOUGHTS

Infusing diversity issues into research methods courses is not only important, it is a fascinating mechanism with which to assist students in thinking critically and creatively about research methods. The methodological challenges associated with research inclusive of diverse populations and the subsequent avoidance of bias may enable students to move away from thinking of research methods as a "canned" science. Instead, students learn to apply what they know about research methods while challenging personal as well as methodological assumptions that may lead to bias, flawed conclusions, or most importantly, the commission of flawed future research.

Ultimately, the continua of humanity extend in many directions, are multifaceted, and infinitely complex. Students are often drawn to psychology because of the puzzle presented by the fascinating diversity of human experience juxtaposed upon remarkable human similarity. We are all very much alike and yet, remarkably unique. This puzzle needs to extend beyond the minds of our students into a psychology curriculum that addresses the diversity of human existence and essence. Psychology as a science of human behavior is incomplete if study samples are primarily composed of individuals fitting certain population parameters. Rather, instructors need to discuss the richness of human diversity, and research needs to be inclusive of the entire spectrum of humanity. As such, it is imperative that future generations of psychologists and researchers responsibly evaluate and conduct research that is both inclusive and reflective of human diversity across the global community.

AUTHORS' NOTE

Linda M. Woolf and Michael R. Hulsizer are in the Department of Behavioral and Social Sciences, Webster University. Correspondence concerning this article should be addressed to: Linda M. Woolf, Department of Behavioral and Social Sciences, Webster University, 470 East Lockwood Ave., Saint Louis, MO 63119, e-mail: woolflm@webster.edu, phone: (314) 968-6970, fax: (314) 963-6094.

REFERENCES

Adair, J. G., Coelho, A. E. L., & Luna, J. R. (2002). How international is psychology? *International Journal of Psychology, 37*, 160–170.

Administration on Aging. (2004). *A profile of older Americans: 2004*. Washington, DC: U. S. Department of Health and Human Services.

American Psychological Association. (2003). Guidelines on multicultural education, training, research, practice, and organizational change for psychologists. *American Psychologist, 58*, 377–402.

Atlani, L., & Rousseau, C. (2000). The politics of culture in humanitarian aid to women refugees who have experienced sexual violence. *Transcultural Psychiatry, 37*, 435–449.

Baltes, P. B. (1968). Longitudinal and cross-sectional sequences in the study of age and generation effects. *Human Development, 11*, 145–171.

Baltes, P. B. (1997). On the incomplete architecture of human ontogeny: Selection, optimization, and compensation as foundation of developmental theory. *American Psychologist, 52*, 366–380.

Beck, A. T., Ward, C. H., Mendelson, M., Mock, J., & Erbaugh, J. (1961). An inventory for measuring depression. *Archives of General Psychiatry, 4*, 561–571.

Bernal, G., Trimble, J. E., Burlew, A. K., & Leong, F. T. L. (2003). Introduction: The psychological study of racial and ethnic minority psychology. In G.

Bernal, J. E. Trimble, A. K. Burlew, & F. T. L. Leong (Eds.), *Handbook of racial and ethnic minority psychology* (pp. 1–12). Thousand Oaks, CA: Sage.

Betancourt, H., & López, S. R. (1993). The study of culture, ethnicity, and race in American psychology. *American Psychologist, 48*, 629–637.

Bowman, S. L. (2003). A call to action in lesbian, gay, and bisexual theory building and research. *The Counseling Psychologist, 31*, 63–69.

Bracken, P. J. (1998). Hidden agendas: Deconstructing post traumatic stress disorder. In P. J. Bracken & C. Petty (Eds.), *Rethinking the trauma of war* (pp. 38–59). New York: Free Association Books.

Brislin, R. (2000). *Understanding culture's influence on behavior* (2nd ed.). New York: Wadsworth.

Buboltz, W. C., Jr., Miller, M., & Williams, D. J. (1999). Content analysis of research in the *Journal of Counseling Psychology* (1973–1998). *Journal of Counseling Psychology, 46*, 496–503.

Butcher, J. N., Graham, J. R., Ben-Porath, Y. S., Tellegen, A., Dahlstrom, W. G., & Kaemmer, B. (2001). *Minnesota Multiphasic Personality Inventory-2 (MMPI-2): Manual for administration, scoring and interpretation (Rev. ed.)*. Minneapolis, MN: University of Minnesota Press.

Carlson, D. K. (2002, February 19). *Acceptance of homosexuality: A youth movement*. Princeton, NJ: The Gallup Organization.

Carter, R. T., Akinsulure-Smith, A. M., Smailes, E. M., & Clauss, C. S. (1998). The status of racial/ethnic research in counseling psychology: Committed or complacent? *Journal of Black Psychology, 24*, 322–334.

Chang, J., & Sue, S. (2005). Culturally sensitive research: Where have we gone wrong and what do we need to do now? In M. G. Constantine & D. W. Sue (Eds.), *Strategies for building multicultural competence in mental health and educational settings* (pp. 229–246). Hoboken, NJ: Wiley.

Council of National Psychological Associations for the Advancement of Ethnic Minority Interests. (2000). *Guidelines for research in ethnic minority communities*. Washington, DC: American Psychological Association.

Delgado-Romero, E. A., Galván, N., Maschino, P., Rowland, M. (2005). Race and ethnicity in empirical counseling and counseling research: A 10-year review. *The Counseling Psychologist, 33*, 419–448.

Denmark, F., Russo, N. F., Frieze, I. R., & Sechzer, J. A. (1988). Guidelines for avoiding sexism in psychological research: A report of the ad hoc committee on nonsexist research. *American Psychologist, 43*, 582–585.

Dresser, R. (2001). Research participants with mental disabilities: The more things change. In L. E. Frost & R. J. Bonnie (Eds.), *Evolution of mental health law* (pp. 57–74). Washington, DC: American Psychological Association.

Dunn, D. S. (2000). Social psychological issues in disability. In R. G. Frank & T. R. Elliott (Eds.), *Handbook of rehabilitation psychology* (pp. 565–584). Washington, DC: American Psychological Association.

Eichler, M. (1991). *Nonsexist research methods: A practical guide*. New York: Routledge.

Elliott, T. R., & Umlauf, R. L. (1995). Measurement of personality and psychopathology following acquired physical disability. In L. A. Cushman & M. J. Scherer (Eds.), *Psychological assessment in medical rehabilitation* (pp. 301–324). Washington, DC: American Psychological Association.

Grace, P. A., Hill, R. J., Johnson, C. W., & Lewis, J. B. (2004). In other words: Queer voices/dissident subjectivities impelling social change. *International Journal of Qualitative Studies in Education, 17*, 301–324.

Graham, S. (1992). "Most of the subjects were White and middle class": Trends in published research on African Americans in selected APA journals, 1970–1989. *American Psychologist, 47*, 629–639.
Guthrie, R. V. (1998). *Even the rat was white: A historical view of psychology* (2nd ed.). Boston: Allyn & Bacon.
Halpern, D. F. (1995). Cognitive gender differences: Why diversity is a critical research issue. In H. Landrine (Ed.), *Bringing cultural diversity to feminist psychology: Theory, research, and practice* (pp. 77–92). Washington, DC: American Psychological Association.
Hames, R., & Draper, P. (2004). Women's work, child care, and helpers-at-the-nest in a hunter-gatherer society. *Human Nature, 15*, 319–341.
Hyde, J. S. (1996). *Half the human experience: The psychology of women* (5th ed.). Lexington, MA: D.C. Heath & Co.
Jones, J. H. (1993). *Bad blood: The Tuskegee syphilis experiment*. New York: Free Press.
Kinsella, K., & Phillips, D. R. (2005). Global aging: The challenge of success. *Population Bulletin, 60*, 3–42.
Koch, N. S., & Emrey, J. A. (2001). The Internet and opinion measurement: Surveying marginalized populations. *Social Science Quarterly, 82*, 131–138.
Kowalski, R. M. (2000). Including gender, race, and ethnicity in psychology content courses. *Teaching of Psychology, 27*, 18–24.
Lerner, R. M. (1995). Developing individuals within changing contexts: Implications of developmental contextualism for human development, research, policy, and programs. In T. J. Kindermann & J. Valsiner (Eds.), *Development of person–context relations* (pp. 13–37). Hillsdale, NJ: Lawrence Erlbaum Associates.
Marsella, A. J., (2001, Spring). Internationalizing the psychology curriculum. *Psychology International, 12*(2), 7–8.
Matsumoto, D. (1994). *Cultural influences on research methods and statistics*. Belmont, CA: Wadsworth.
Matsumoto, D. (2003). Cross-cultural research. In S. F. Davis (Ed.), *Handbook of research methods in experimental psychology* (pp. 189–208). Malden, MA: Blackwell.
McHugh, M. C., & Cosgrove, L. (2002). Gendered subjects in psychology: Satirical and dialectic perspectives. In L. H. Collins, M. R. Dunlap, & J. C. Chrisler (Eds.), *Charting a new course for feminist psychology* (pp. 3–19). Westport, CT: Praeger.
McHugh, M. C., Koeske, R. D., & Frieze, I. H. (1986). Issues to consider in constructing nonsexist psychological research: A guide for researchers. *American Psychologist, 41*, 879–890.
Morrow, S. L. (2003). Can the master's tools ever dismantle the master's house? Answering silences with alternative paradigms and methods. *The Counseling Psychologist, 31*, 70–77.
Munley, P. H., Anderson, M. Z., Baines, T. C., Borgman, A. L., Briggs, D., Dolan, J. P., Jr., et al. (2002). Personal dimensions of identity and empirical research in APA journals. *Cultural Diversity and Ethnic Minority Psychology, 8*, 357–365.
Pate, W. E., II. (2001). *Analyses of data from graduate study in psychology: 1999-2000*. Washington, DC: American Psychological Association. Retrieved October 13, 2005, from http://research.apa.org/grad00contents.html
Pullin, D. (2002). Testing individuals with disabilities: Reconciling social science and social policy. In R. B. Ekstrom, & D. K. Smith (Eds.), *Assessing individuals*

with disabilities in educational, employment, and counseling settings (pp. 11–31). Washington, DC: American Psychological Association.

Quina, K., & Bronstein, P. (2003). Gender and multiculturalism: Transformations and new directions. In P. Bronstein & K. Quina (Eds.), *Teaching gender and multicultural awareness: Resources for the psychology classroom* (pp. 3–1).Washington, DC: American Psychological Association.

Quina, K., & Kulberg, J. M. (2003). The experimental psychology course. In P. Bronstein & K. Quina (Eds.), *Teaching gender and multicultural awareness: Resources for the psychology classroom* (pp. 87–98).Washington, DC: American Psychological Association.

Rochlin, M. (1995). The Heterosexual Questionnaire. In M. S. Kimmel & M. A. Messner (Eds.), *Men's lives* (3rd ed., p. 405). Boston: Allyn & Bacon.

Rosenberg, N. A., Pritchard, J. K., Weber, J. L., Cann, H. M., Kidd, K. K., Zhivotovsky, L. A., et al. (2002). Genetic structure of human populations. *Science, 298*, 2381–2385.

Saad, L. (2005, May 20). *Gay rights attitudes a mixed bag: Broad support for equal job rights, but not for gay marriage*. Princeton, NJ: The Gallup Organization.

Savin-Williams, R. C. (2001). *Mom, Dad. I'm gay. How families negotiate coming out*. Washington, DC: American Psychological Association.

Schaie, K. W. (1986). Beyond calendar definitions of age, time, and cohort: The general developmental model revisited. *Developmental Review, 6*, 252–277.

Schaie, K. W. (2000). The impact of longitudinal studies on understanding development from young adulthood to old age. *International Journal of Behavioral Development, 24*, 257–266.

Sears, D. O. (1986). College sophomores in the laboratory: Influences of a narrow data base on social psychology's view of human nature. *Journal of Personality and Social Psychology, 51*, 515–530.

Shin, H. B. (2005). *School enrollment—Social and economic characteristics of students: October 2003*. (Pub. No.P20-554). Washington, DC: U. S. Census Bureau.

Sherif, C. W. (1979). Bias in psychology. In J. Sherman & E. T. Beck (Eds.), *The prism of sex: Essays in the sociology of knowledge* (pp. 107–146). Madison: The University of Wisconsin Press.

Steele, C. M., Spencer, S. J., & Aronson, J. (2002). Contending with group image: The psychology of stereotype and social identity threat. In M. Zanna (Ed.), *Advances in experimental social psychology*, (Vol. 34, pp. 379–440). San Diego, CA: Academic Press.

Sue, S. (1999). Science, ethnicity, and bias: Where have we gone wrong? *American Psychologist, 54*, 1070–1077.

United States Census Bureau. (2003). *2003 American community survey*. Retrieved October 13, 2005, from http://www.census.gov/acs/www/index.html

United States Census Bureau. (2004). *Projected population of the United States, by race and Hispanic Origin: 2000 to 2050*. Retrieved October 13, 2005, from http://www.census.gov/ipc/www/usinterimproj/

Wang, V. O., & Sue, S. (2005). In the eye of the storm: Race and genomics in research and practice. American Psychologist, *60*, 37–45.

Weisstein, N. (1971). Psychology constructs the female, or the fantasy life of the male psychologist (with some attention to the fantasies of his friends, the male biologist and the male anthropologist). *Social Education, 35*, 362–373.

Woolf, L. M., Hulsizer, M. R., & McCarthy, T. (2002). *International psychology: A compendium of textbooks for selected courses evaluated for international content*.

Office of Teaching Resources in Psychology. Retrieved October 13, 2005, from http://www.lemoyne.edu/OTRP/otrpresources/inter-comp.html

Yan, E. G., & Munir, K. M. (2004). Regulatory and ethical principles in research involving children and individuals with developmental disabilities. *Ethics & Behavior, 14,* 31–49.

Yoder, J. D., & Kahn, A. S. (1993). Working toward an inclusive psychology of women. *American Psychologist, 48,* 846–850.

Zuckerman, M. (1990). Some dubious premises in research and theory on racial differences. *American Psychologist, 45,* 1297–1303.

Chapter **17**

Teaching Writing in Statistics and Research Methods: Addressing Objectives, Intensive Issues, and Style

Michelle E. Schmidt and Dana S. Dunn
Moravian College

Consider Freud, Lewin, and Skinner—among the most important figures in psychology. Some of their papers are awkward, incoherent, and inconsequential. But we honor their good papers. I am sure that none of these individuals set out to write classic papers. They simply wrote.

—Christopher Peterson, 1996, p. 284

One aphorism for writers is this: Writers write. This maxim holds true equally for late great psychologists, professional writers and researchers, and would-be student authors in psychology. As Peterson (1996) implied, neither divine inspiration nor a muse are necessary to create serviceable prose. For students, especially, the important issue is apt to be learning to write using the vernacular and grammar of our discipline, as there is professional as well as pedagogical substance to it (Madigan, Johnson, & Linton, 1995).

This chapter addresses the importance of writing as part of the teaching and learning of statistics and research methods in the psychology curriculum. We present (a) research that suggests that writing within psychology and, specifically, within statistics and methods courses is critical; (b) our model for teaching a writing-intensive research methods and statistics course; and (c) assignments that others find useful to teach writing in these courses.

THE IMPORTANCE OF WRITING

The Writing Across the Curriculum (WAC) movement was born in the 1970s (Russell, 1992). Mid-decade, the National Assessment of Educational Progress revealed that writing skills among students were declining; American educational institutions did not adequately teach their students to write. This discovery fueled interest in the WAC model, and colleges and universities across the country began working under the assumption that students could learn fundamentals of writing outside of traditional composition courses taught primarily by English department faculty.

Similar to professionals in other disciplines, teachers in psychology have become increasingly interested in teaching effective and appropriate writing to their students. Recently, a task force study endorsed by the American Psychological Association (APA) underscored the importance of statistics and research methods and the communication of the information learned in this course (available online at www.apa.org/ed/pcue/taskforcereport.pdf). The *Undergraduate Psychology Major Learning Goals and Outcomes: A Report*, lists research methods and communication skills as 2 of the 10 identified goals. The report states, "Students will understand and apply basic research methods in psychology, including research design, data analysis, and interpretation ... Students will be able to communicate effectively in a variety of formats." Thus, it is important for instructors of psychology to design assignments that teach students the skills they need to share their newly acquired knowledge about statistics and research methods.

Despite some social science faculty members' resistance to writing-intensive courses (e.g., extra faculty workload, insufficient class time, student dissatisfaction, faculty feelings of inadequacy with regard to teaching writing, instructors' personal distaste for writing; Boice, 1990), studies of pedagogy demonstrate that a focus on writing in statistics and research methods is associated with better learning. For example, in an empirical study of the effectiveness of writing assignments on comprehension of statistics, Beins (1993) found that a greater concentration on writing was associated with better mastery of the course material. Additionally, in a study of innovative teaching techniques, which included, for example, regular essays written in the form of letters about various statistical concepts and journal article critiques, Pan and Tang (2004) reported that these techniques were associated with lower anxiety about statistics and, presumably, with better learning.

A MODEL FOR A WRITING INTENSIVE RESEARCH METHODS AND STATISTICS COURSE

Writing Intensive (WI) courses at Moravian College, a selective liberal arts institution, fulfill a WAC requirement. After completing a required Writing 100 course taught by faculty from across the college, students are required to take a WI course within their major. In the spirit of WAC, WI courses must adhere to the following guidelines: (a) There must be opportunities for the writing process to work (e.g., feedback on drafts); (b) instructors must use some informal or exploratory—and ungraded—writing assignments (e.g., freewriting; Belanoff, Elbow, & Fontaine, 1991); (c) at least 50% of the course grade must be based on assessment instruments that employ written discourse (e.g., exams, papers, labs); (d) a minimum of 25 pages of writing must be completed by the end of the course; and (e) there must be practice in the conventions of writing within the course's discipline (e.g., MLA style, APA style).

Within our six-member department, we have a two-course sequence for statistics and methods (Psychology 211–212: Experimental Methods and Data Analysis I and II). Although the first course in the sequence serves as the designated WI course for our department, both courses meet the requirements of WI. The main course objectives for the two-course sequence are (a) learning statistical techniques for interpreting psychological data and (b) learning to write well within the discipline using APA style. In order to meet these two objectives, students design a study and write a corresponding proposal that reviews relevant literature, provide a rationale for their study, and outline the design and methodology of the proposed study (Psychology 211). Students then carry out their study by recruiting participants and collecting data, analyzing their data according to the hypotheses in the proposal, and completing the full APA style paper with an introduction and Method section (taken from the proposal) and Results and Discussion sections (Psychology 212; sample course syllabi can be found on the CD accompanying this volume).

PRIMARY WRITING ASSIGNMENTS FOR STATISTICS AND METHODS

The requirements for most statistics and research methods courses, including the tasks outlined previously in our two-course sequence, are daunting for most students. In particular, statistics anxiety is negatively associated with learning (Onwuegbuzie & Seaman, as cited in Pan & Tang, 2004). Some researchers have gone so far as to liken learning statistics to learning a foreign language (e.g., Lalande & Gardner, cited in Pan & Tang, 2004). Due to the heavy and intimidating workload and the high levels of anxiety felt by students in statistics courses, we believe that it is important to reduce the APA-style research paper to a series of manageable smaller tasks. We present the writing assignments that we find useful subsequently (see Table 17.1). We break assignments into two categories: (a) those that introduce the students to APA style and (b) those that are stepping-stones to their research proposal and their eventual full APA-style research paper.

TABLE 17.1
Assignments in Moravian College's Model of Statistics and Research Methods

Learning APA Style Assignments	*Graded? (Y/N)*
Reference page assignment	Y
Article critique	
Group lab assignment	N
Individual critiques	Y
Weekly lab assignments	Y
Stepping-Stone Assignments	
Three things I'm interested in	N
How I would research my first choice	N
Draft of introduction and method section with annotated bibliography	N
Human Subjects Internal Review Board proposal	Y
Conference proposal	Y
Full paper draft	N
Conference poster/paper presentation	Y
Final APA style research paper	Y

Learning APA-Style Conventions

The Reference Page. We give students a packet of APA-style Reference pages that have a variety of style errors. Students must correct the errors on the form using the guidelines outlined in the *APA Publication Manual* (APA, 2001). We find that this assignment gets students to pay attention to the subtleties of APA-style citations.

Article Critique. Although not directly contributing to their research proposal, students complete an article critique on the introduction and Method sections of an unpublished manuscript. Students complete one critique on an instructor-provided article in small groups during lab time. As a class, the instructor and students discuss the article and the groups' critiques of the article. Students then individually write a critique on a second article provided by the instructor. Critiques explore questions outlined in the *Publication Manual* (2001). For example, "Is the research question significant, and is the work original and important?" (p. 6). "How do the hypothesis and the experimental design relate to the problem?" (p. 16; see Dunn, 2004, for more detailed directions for assigning and crafting article critiques). These questions familiarize students with the first two sections of the APA paper and get them used to using the *APA Manual*.

Weekly Lab Assignments. With each chapter of the text, students complete a laboratory assignment in which they practice analyzing data with a statistical test. They complete analyses by hand first and then confirm the results using SPSS. Students then write a short introduction on the topic, provide a rationale for the study, and present the hypothesis(es). Next, they write a Method section, using basic information provided by the instructor. As the semester progresses and students gain more skills, these lab reports include Results and Discussion sections, which become progressively more detailed.

Stepping Stones

Three Things I'm Interested In. We begin the semester by instructing students to identify topics that they might be interested in studying empirically. We encourage them to think back to introductory psychology and choose topics that are researchable and appealing to them. They must identify the independent and dependent variables for their potential study and list keywords that they would use to do a literature search for each topic. To prepare students to do literature searches, we bring in a reference librarian who teaches the students how to use the database for articles in psychology. The librarians show them how to access the database, thoroughly search for articles, and acquire the articles themselves.

How I Would Research My First Choice (With Annotated Bibliography). After receiving feedback on their suggested topics of interest, students begin to research one in earnest. The annotated bibliography exercise provides students the opportunity to work on APA style for the Reference page and write summaries of the located articles. Through this process, the instructor can judge APA style competency and provide students with feedback on whether their articles will likely be appropriate for the introduction of their paper.

In a similar vein, Henderson (2000) described an excellent project for students to review the psychological literature on a topic while preparing to write an APA-style paper or to conduct a lab. The project is a "reader's guide," a synopsis that provides (a) a content outline (e.g., historical background, theoretical and methodological issues, major issues and research areas); (b) theorists and contributors (a list of the main workers in an area, accompanied by brief, three-sentence characterizations explaining their role and importance); (c) central concepts (a list of the essential 10 or so concepts in an area);(d) hot or current topics (what topics recur, what cause debate); and (e) major resources (important chapters, handbooks, chapters in the *Annual Review of Psychology*, main journals, Internet sources). Students learn to be critical readers as they sift through the relevant literature in order to identify material that will contribute to their reader's guides.

Draft of Introduction and Method Section. After students have had sufficient time to learn about the construction of the Introduction and Method section of the APA paper, students present a draft of their research proposal. It is often helpful to refer students to practical guides to APA style for additional pointers on writing with appropriate style (e.g., Dunn, 2004; Mitchell, Jolley, & O'Shea, 2004; Sternberg, 1989; Szuchman, 1999). The student research proposal includes an Abstract, introduction, Method section, and Reference page. The instructor reviews the draft and provides the student with ungraded feedback. We find that ungraded feedback helps students to get past their difficulty with critical feedback (Nadelman, cited in Goddard, 2003; see also Elbow, 1993). One draft is required, but students may continue to get ungraded feedback on subsequent drafts, if they choose to do so. Students turn in the final draft of the proposal at the end of the first course. The proposal becomes the starting point for the second semester course. (For pedagogical continuity, the same instructor teaches both courses with the same group of students across the academic year.)

HSIRB Proposal. The second course in the sequence focuses on data collection, analyses, and interpretation of results. Before collecting data, students submit their research proposal to the Human Subjects Internal Review Board (HSIRB). At Moravian College, the HSIRB committee contains a group of eight faculty from different disciplines. The proposal includes information about the study—rationale, focus, participants, measures, and ethical considerations. HSIRB members review the proposals and request revisions to the methodology where necessary. Beyond getting institutional approval for the study on ethical grounds, this exercise offers students an opportunity to write about their study to an audience of individuals outside of the discipline. They must write in clear prose, and their rationale for the study must be coherent, thoughtful, and convincing.

Conference Proposal. A small local conference provides an opportunity for students to write a conference proposal and ultimately share their research with others. The consortium of six colleges and universities in our area holds an annual psychology undergraduate research conference (see http://www.lvaic.org for the announcement of this annual conference and call for papers). Most faculty in our department now require their students to participate in this conference. The typical proposal format is a 250- to 500-word abstract of the student's work.

Full Paper Draft. Students modify the Introduction and Method section of their research proposal and add results and Discussion sections for the full APA research report. Students also add an Abstract and appropriate tables and figures to the draft of the final paper. The instructor again provides feedback on the draft(s). Some instructors also encourage students to share working drafts with one another either during an in-class workshop or outside the classroom (Dunn, 1994). By this time in the semester, students usually feel comfortable writing these sections of the paper because they have had a number of lab assignments that gave them practice.

Conference Poster/Paper Presentation. Students present their research findings in either paper or poster form at the local conference. Once again, students have another opportunity to write about their study and communicate the information in yet another format. This assignment adds a visual element to the process: Students must create data displays—tables and figures—that both support and that link back to their writing. Of course, the verbal element remains of paramount importance, especially because transforming a paper into a poster requires that students learn to condense their writing and to highlight only the essential issues (Dunn, 2004). What better way to teach students to get the heart of some matter, the real meaning of their work, than to turn a 15- to 20-page paper into a brief, focused poster? Students must understand what they wrote, to the point that they can answer questions about their study as peers mill about their poster.

Final APA-style Research Paper. The final APA research paper, due at the end of the second semester of our integrated course, is a full manuscript that should effectively communicate the findings of the student's original study. By the academic year's end, students have had multiple drafting opportunities, received copious ungraded feedback, and completed many smaller practice assignments. The expectation is that, by this point, they should be proficient in communicating information about research methods and statistics through their writing.

EVALUATING OUR WRITING MODEL FOR STATISTICS AND METHODS

The list of assignments that we used in our two-course sequence for statistics and methods provides a WI model for teaching APA style. The model presents students with writing as a process, opportunities for improvement without punishment (i.e., ungraded feedback, mastery learning approaches), and multiple assignments from which they can learn the specifics of APA style.

There are a number of benefits to this intensive approach. When we test students on parts of the APA research paper, they tend to remember what information resides in which section of the paper and how to interpret and communicate statistical information. Additionally, the APA research paper no longer seems like something they cannot do. Students are more at ease with the process of writing and seem to like the process more.

The model that we propose is demanding for students and instructor alike. Students must dedicate a great deal of time to writing, and the instructor must spend a great deal of time reading students' work in order to provide constructive feedback. Overall, however, the time put into the work appears to be associated with better learning as demonstrated by exams and final written products. In fact, on student evaluations, it is not uncommon for students to recommend still more opportunities for drafting the final paper.

Student Evaluations

Anecdotal evidence in the form of testimonials indicates that our students are pleased to learn about writing in psychology gradually, across an entire academic year rather one semester. Students especially welcome the opportunity to craft an APA-style paper summarizing a research project in stages. Receiving in-depth comments from the instructor and peers in the class allows students to expand their working knowledge of APA style in phases.

In addition to the traditional student evaluation items, students rated additional items specific to the writing aspect of the course. Student feedback was very positive. For example, student evaluations contained an item that stated, "Writing papers in multiple drafts was useful to me." Using a scale of 1 (*strongly disagree*) to 5 (*strongly agree*), the mean rating by students in the class ($N = 16$) was 4.25 ($SD = .68$). Similarly, in response to the evaluation item, "The writing assignments in the course helped me to learn the material content of the course," students gave a mean rating of 4.25 ($SD = .62$). Finally, students appreciated that the writing in the course taught them about writing in the discipline, more generally. In response to the evaluation item, "The writing assignments in this course helped me to learn and practice the conventions of writing common to the discipline of which the course is part," students gave a mean rating of 4.44 ($SD = .63$).

In the narrative portion of the evaluations, students have the opportunity to comment on which aspects of the course were helpful and should be continued in future semesters. One student noted, "I liked all of the mini [assignments] that were due for the proposal instead of just having a paper due at the end [of the semester]. I like the fact that we handed in rough drafts of the proposal and [they were] ungraded." Another student used both drafting and the article critique assignments as examples of how repeated opportunities to learn APA style were helpful.

Assessment Issues in Student Writing

The growing movement toward demonstrating accountability in higher education spawned widespread interest in academic assessment (e.g., Astin, 1993; Maki, 2001). Psychologists have not been idle bystanders: Various works addressing the importance of assessment—measuring and evaluating student learning in the discipline—are now available (e.g., Dunn, Mehrotra, & Halonen, 2004; Halonen et al., 2003; Halpern et al.,

1993). Within psychology, the issue of assessing the quality and depth of student writing has received less emphasis than other areas of assessment, but this, too, is changing (Stoloff, Apple, Barron, Reis-Bergan, & Sundre, 2004; Stoloff & Rogers, 2002).

Indeed, Halonen et al.'s (2003) rubric for teaching, learning, and assessing scientific reasoning in psychology points to writing as a cornerstone within the communications skills domain (see also Bosack, McCarthy, Halonen, & Clay, 2004). The rubric outlines levels of expected communication proficiency prior to training through the professional graduate level and beyond. Of interest to instructors teaching writing are students' resource-gathering skills, argumentation skills, and use of conventional expression (including APA format) at the developing and advanced undergraduate levels. Teachers can use the rubric to plan the content of their courses and to identify the core writing and reasoning abilities students should display, including evaluating scholarly evidence, organization, awareness of audience, persuasiveness, and grammar and punctuation. A side benefit of successfully teaching students about writing in psychology is that they should develop self-assessment competencies—reflection and regulation—in this domain (e.g., Dunn, McEntarffer, & Halonen, 2004). That is, students should be able to accomplish tasks associated with the writing and research process and to critique their progress with the goal of self-improvement in the future (see also, Halonen et al.'s comments on the self-assessment skills domain).

OTHER EFFECTIVE WRITING ASSIGNMENTS IN STATISTICS AND METHODS COURSES

The previous writing assignments and process are helpful to students as they try to learn statistics and research methods and the appropriate ways to communicate this information. These assignments could be taken out of the larger picture and used a la carte. In fact, others have described a number of innovative writing assignments (see Table 17.2).

Faculty can encourage students to write in a statistics course in a variety of ways. In this section, we summarize writing techniques that also should be useful to instructors of statistics and research methods. Three categories summarize writing assignments according to instructors' learning goals for students: (a) enhanced understanding of statistical concepts, (b) mastery of APA style, and (c) self-evaluation of the student's learning.

Enhanced Understanding of Statistical Concepts

One of the fundamental learning goals in teaching statistics is teaching students to transfer knowledge of when and how to apply a test statistic to data and to then explain in prose form what a result means. Learning to mix the practice of data analysis with the conceptual side of interpreta-

TABLE 17.2
Other Writing Assignments

Assignment	*Source*
Reader's guide	Henderson (2000)
In-class IRB	Dunn (1999)
Practical books on APA-style writing	Dunn (2004)
	Mitchell, et al. (2004)
	Sternberg (1989)
	Szuchman (1999)
Turning statistical results into prose	Dunn (1999, 2004)
Interpret generated statistical output	Dolinsky (2001)
	Dunn (1999, 2001, 2004)
	Smith (1995)
Press release	Beins (1993)
Ungraded writing to learn assignments	Nodine (1990)
Five assignments for learning to writing a literature review	Poe (1990)
Collaborative writing	Dunn (1996)
	Nodine (1990)
Student publications and APA-style mechanics	Ware, et al. (2002)
Learning assessment journal	Dolinsky (2001)
Letter writing	Dunn (2000)
	Keith (1999)
	Pan & Tang (2004)
Research methods script	Wilson & Hershey (2000)
Freewriting and peer editing	Dunn (1994)
Portfolios	Rickabaugh (1993)
General writing course for the major	Goddard (2002, 2003)

tion is not always easy. In fact, it is not difficult to argue that the failure of knowledge transfer from one context to the other is the root learning problem for many psychology students—applying a statistical test to a book example is a far cry from doing so with a novel data set contrived by an instructor or the observations from one's own experiment.

Dunn (1999, 2004; see also, Dunn, 2001) encouraged students to link the purpose of a statistical test with a prose outcome by completing flowcharts linking the two together. Students begin the process by reflecting on the properties of a given test statistic (e.g., the independent groups t test) and its use (i.e., comparing means from two different groups), performing an analysis (presumably using statistical software), and identifying and reviewing the statistical notation (i.e., degrees of freedom, value of the test statistic, critical value). They then review the study's hypothesis in conceptual (the theory being tested) and practical terms (the operationalization of the theory). Finally, students write about the observed results using numerical information (e.g., means, statistical notation) with explicit descriptions of behavior (i.e., what participants did or did not do behaviorally). When appropriate, students create APA-style tables or figures, linking these visual displays to the text by leading the reader through the data.

A number of assignments encourage students to think about the statistics that they generate. As we mentioned previously, an assignment that gives students practice writing involves simply requiring students to interpret the meaning of generated statistical output (Dolinsky, 2001). Smith (1995) described open-ended assignments in which students generate a research question, choose the appropriate statistics to run, and then answer their question using their generated output. These additions to the traditional "compute the numbers" assignment provide students with additional opportunities to develop an understanding of the statistics that they generate. Students also can write a short press release to communicate statistical findings to a lay audience (Beins, 1993). This interesting assignment encourages understanding of statistical content as students learn important interpretive skills.

Nodine (1990) advocated that teachers assign ungraded, in-class, writing-to-learn exercises that help students make sense of data. Students can work individually, in pairs, or small groups, sharing their writing during class discussion. Instructors need only provide students with figures, line graphs, or bar graphs illustrating some numerical relationship among variables. The students are responsible for writing clear, descriptive legends that explain the relationships illustrated in the graphs or figures. Instructors also can present students with a table of numerical data, inviting them to write a description of what information the table shows. Alternatively, Nodine noted that she occasionally gives the students a written description of a graph's legend, inviting them to draw an accompanying visual representation for it.

Mastery of APA Style

Most instructors in psychology-based statistics and methods courses require their students to learn APA style. So that students will be successful at writing literature reviews, Poe (1990) taught students useful skills through a series of five assignments. First, students freewrite about three topics of interest to them. Second, the instructor discusses with students how to write good (and bad) article summaries, and students summarize an instructor-provided essay (plagiarism is also addressed). Third, students read a journal article with the abstract removed and write an abstract for the article. Students then compare their abstract to the published abstract. Fourth, students repeat the abstract writing assignment, this time with instructor feedback. Fifth, students write a statement about their research topic and provide a reference list with 10 sources to support the topic. Often we teach our students what goes into each section of the APA paper, but we do not offer them this more micro level of guidance and feedback. This series of assignments shapes the students so that they are able to write the introduction without the instructor's assumption that they should be able to summarize articles.

In a similar style to Poe (1990), Dunn (1996) described collaborative writing as a learning tool in a statistics and research methods course. Paired students work together throughout the semester to create, design, conduct, and report on an experiment. First, students write a prospectus through a shared writing process (shared prewriting, freewriting, outlining, drafting, and revising). Second, together, students use the instructor feedback on the prospectus to work toward a final research paper. Third, a peer-review workshop takes place during class time. Each pair of students works with another pair in a paper exchange. Students use feedback skills learned in class to give detailed feedback to each other. Students make final edits accordingly. The emphasis in this course is collaborative writing as a tool for learning APA style research papers. Interestingly, 96% of the students in this course believed that they learned more in the class due to the collaborations. (For a related example of collaborative writing, see Nodine, 1990; 2002.)

Ware, Badura, and Davis (2002) recommend using student scholarship—research articles in psychology written by undergraduate students and published in journals devoted to undergraduate research—as a pedagogical device for mastering APA-style mechanics. The authors described writing exercises based on having students read published peer articles in or outside the classroom. After reading an article, students can search for acceptable and unacceptable examples of the passive voice, determine whether the article's title and abstract adhere to APA length restrictions, compare citations in the text with the actual References list (i.e., they should match). Ware and colleagues use these and other activities to demonstrate to students that although errors appear in published works, careful reading of such works develops research and writing skills, as well as critical thinking and active in-class engagement.

Self-Evaluation of Student Learning

In addition to teaching course content, some argue that students benefit from taking time to reflect on the learning process. For example, Dolinsky (2001) reported that she requires students to keep biweekly journals in which they write on a general topic provided by the instructor (e.g., feelings toward statistics). The goals of this assignment are threefold: provide information to the instructor on the students' perceptions of learning, give students an opportunity to think about their learning, and inform the instructor on students' mastery of topics in the course. The journal becomes a tool for both the student and the instructor.

A letter-writing exercise created by Dunn (2000) also promoted a bit of self-reflection and self-assessment of learning where statistics are concerned. Dunn had students in separate sections of a research methods and statistics course write letters back and forth to one another. Students identified statistical or methodological concepts they found to be interesting, challenging, or otherwise noteworthy. Many students used the letters as an opportunity to disclose what they learned about the properties of a particular test statistic (e.g., comparison of more than two means in the analysis of variance) or the essential nature of randomization in psychological experimentation. After sending a letter to a peer in the other section, each student received a letter, read it, and replied. The response letters were aimed at affirming the letter writers' choice of topics while giving the respondents a chance to reflect on their peers' reactions. Several students elected to write a third (unassigned) letter back to their peer correspondents. (For other examples of letter writing exercises, see Keith, 1999; Pan & Tang, 2004.)

Wilson and Hershey (1996) described an exercise in which students identify the concrete steps (in script form) psychologists take between developing an idea for a research project and then publishing a paper based on the idea. Students can do this exercise in class in 10 or so minutes and then share the steps they identified through group discussion. Variations include having students work in pairs or having students generate a script for a research topic for which they will subsequently collect data. The original goal of the exercise was to allow students (novices) to compare their research planning with that of experts (research psychologists). Students can clearly benefit from thinking through a piece of research from start to finish.

Other General Writing Exercises

Regardless of the specific assignments once chooses, we must remember that writing assignments should be designed to help students write well, not simply to write using a designated style (Dunn, 1994). There are useful writing exercises that others have used in courses outside of statistics and methods that can be readily adapted in statistics and methods courses. For example, freewriting and peer editing are techniques that can be useful throughout the APA research paper writing process (Dunn, 1994). When students first try to find a topic of interest to them, they might try freewriting in order to bring their ideas together. In courses where students use different data sets for their APA research paper, peer editing can be helpful.

Instructors use portfolios both inside and outside of psychology courses. In order to reduce writing anxiety, Rickabaugh (1993) created a portfolio made up of small, self-directed laboratory type assignments that students completed outside of class time. Although Rickabaugh did not use this portfolio assignment in a statistics and methods course, it certainly could be easily adapted into the course. Students could complete laboratory assignments and collect them in a portfolio, thus creating a useful study tool for the semester and encouraging students to write. Students also could collect smaller APA-style learning assignments in a portfolio that could serve as a collection of APA-style examples to refer to later in the course or in subsequent courses.

In some departments, good writing in statistics and research methods courses may need supplementing by other writing opportunities. Goddard (2003) taught a writing course for psychology majors. Some students took the course before their statistics and methods course and others did not. As one would expect, those who took the general writing course before statistics and methods found statistics and methods easier, and those who had not taken statistics and methods had more difficulty with the course. A lesson, perhaps, from this writing course model is that a course that teaches writing for the major may be a helpful precursor to the traditional statistics and methods course. With the many requirements in statistics and methods courses, taking some of the stress out of learning APA style might be helpful.

Of course, a caveat to too much focus on learning APA style should be added to any stand-alone course on writing in psychology. Students need to learn to write in general—and not exclusively in the language and folkways of academic psychology. To combat overemphasizing the traditional APA-style lab report in an upper level seminar titled Writing in Psychology, the second author required students to do various types of writing, including short and long book reviews, an op-ed piece, a literature review (in the fashion of a *Psychological Bulletin* article), a poster and a conference paper, and a semester-long group project. (A copy of the syllabus can be found on the CD accompanying this volume.) Students still wrote an APA-style lab report, but when they completed the course, they had a portfolio of additional writings in psychology. (For further thoughts on teaching a writing course in psychology, see Goddard, 2002).

CLOSING THOUGHTS

> *Writing is often spoken of in terms of art, inspiration, genius, and the like. Writing in actuality is a mundane activity that takes place over time. Repetitive actions are the essence of writing.*
>
> —Christopher Peterson, 1996, p. 282

Statistics and methods courses for the psychology major tend to evoke a great deal of stress and anxiety. How can instructors reduce the fear factor while encouraging students to realize the importance of writing and teaching them to write well? As old fashioned as it may seem, salvation may be found in repetition: Writers write. Student writers especially must write—and as often as they possibly can. We encourage teachers and their students to keep these words of wisdom in mind.

ACKNOWLEDGMENT

Portions of this work were undertaken when a 2005 Moravian College Faculty Development and Research Committee Summer Stipend supported the second author.

REFERENCES

American Psychological Association. (2001). *Publication manual of the American Psychological Association* (5th ed.). Washington, DC: Author.

Astin, A W. (1993). *Assessment for excellence: The philosophy and practices of assessment and evaluation in higher education.* New York: Macmillan.

Beins, B. C. (1993). Writing assignments in statistics classes encourage students to learn more interpretation. *Teaching of Psychology, 20,* 161–164.

Belanoff, P., Elbow, P., & Fontaine, S. I. (Eds.). (1991). *Nothing begins with n: New explorations of freewriting.* Carbondale and Edwardsville: Southern Illinois University Press.

Boice, R. (1990). Faculty resistance to writing-intensive courses. *Teaching of Psychology, 17,* 13–17.

Bosack, T. N., McCarthy, M. A., Halonen, J. S., & Clay, S. P. (2004). Developing scientific inquiry skills in psychology: Using authentic assessment strategies. In D. S. Dunn, C. M. Mehrotra, & J. S. Halonen (Eds.), *Measuring up: Educational assessment challenges and practices for psychology* (pp. 141–169). Washington, DC: American Psychological Association.

Dolinksy, B. (2001). An active learning approach to teaching statistics. *Teaching of Psychology, 28,* 55–58.

Dunn, D. S. (1994). Lessons learned from an interdisciplinary writing course: Implications for student writing in psychology. *Teaching of Psychology, 21,* 223–227.

Dunn, D. S. (1996). Collaborative writing in a statistics and research methods course. *Teaching of Psychology, 23,* 38–40.

Dunn, D. S. (1999). *The practical researcher: A student guide to conducting psychological research.* New York: McGraw-Hill.

Dunn, D. S. (2000). Letter exchanges on statistics and research methods: Writing, responding, and learning. *Teaching of Psychology, 27,* 128–130.

Dunn, D. S. (2001). *Statistics and data analysis for the behavioral sciences.* New York: McGraw-Hill.

Dunn, D. S. (2004). *A short guide to writing about psychology.* New York: Pearson Longman.

Dunn, D. S., McEntarffer, R., & Halonen, J. S. (2004). Empowering psychology students through self-assessment. In D. S. Dunn, C. M. Mehrotra, & J. S. Halonen (Eds.), *Measuring up: Educational assessment challenges and practices for psychology* (pp. 171–186). Washington, DC: American Psychological Association.

Dunn, D. S., Mehrotra, C. M., & Halonen, J. S. (Eds.). (2004). *Measuring up: Educational assessment challenges and practices for psychology.* Washington, DC: American Psychological Association.

Elbow, P. (1993). Ranking, evaluating, and liking: Sorting out three forms of judgment. *College English, 55,* 187–206.
Goddard, P. (2002). Promoting writing among psychology students and faculty: An interview with Dana S. Dunn. *Teaching of Psychology, 29,* 331–336.
Goddard, P. (2003). Implementing and evaluating a writing course for psychology majors. *Teaching of Psychology,* 30, 25–29.
Halonen, J. S., Bosack, T., Clay, S., & McCarthy, M. (with Dunn, D. S., Hill IV, G. W., McEntarffer, R., Mehrotra, C. M., Nesmith, R., Weaver, K., & Whitlock, K.). (2003). A rubric for authentically learning, teaching, and assessing scientific reasoning in psychology. *Teaching of Psychology, 30,* 196–208.
Halpern, D. F., Appleby, D. C., Beers, S. E., Cowan, C. L., Furedy, J. J., Halonen, J. S., Horton, C. P., Peden, B. F., & Pittengeret, D. J. (1993). Targeting outcomes: Covering your assessment concerns and needs. In T. V. McGovern (Ed.), *Handbook for enhancing undergraduate education in psychology* (pp. 23–46). Washington, DC: American Psychological Association.
Henderson, B. B. (2000). The reader's guide as an integrative writing experience. *Teaching of Psychology, 28,* 257–259.
Keith, K. D. (1999). Letters home: Writing for understanding in introductory psychology. In L. T. Benjamin, B. F. Nodine, R. M. Ernst, & C. Blair Broeker (Eds.), *Activities handbook for the teaching of psychology* (Vol. 4, pp. 30–32). Washington, DC: American Psychological Association.
Madigan, R., Johnson, S., & Linton, P. (1995). The language of psychology: APA style as epistemology. *American Psychologist, 50,* 428–436.
Maki, P. L. (2001). From standardized tests to alternative methods: Some current resources on methods to assess learning in general education. *Change, 33*(2), 29–31.
Mitchell, M. L., Jolley, J. M., & O'Shea, R. P. (2004). *Writing for psychology.* Belmont, CA: Wadsworth.
Nodine, B. F. (1990). Assignments in psychology: Writing to learn. In T. Fulwiler & A. Young (Eds.), *Programs that work: Models and methods for writing across the curriculum* (pp. 146–148). Portsmouth, NH: Boynton/Cook.
Nodine, B. F. (2002). Writing models, examples, teaching advice, and a heartfelt plea. In S. F. Davis & W. Buskist (Eds.), *The teaching of psychology: Essays in honor of Wilbert J. McKeachie & Charles L. Brewer* (pp. 107–120). Mahwah, NJ: Lawrence Erlbaum Associates.
Pan, W., & Tang, M. (2004). Examining the effectiveness of innovative instructional methods on reducing statistics anxiety for graduate students in the social sciences. *Journal of Instructional Psychology, 31,* 149–159.
Peterson, C. (1996). Writing rough drafts. In F. T. L. Leong & J. T. Austin (Eds.), *The psychology research handbook: A guide for graduate students and research assistants* (pp. 282–290). Thousand Oaks, CA: Sage.
Poe, R. E. (1990). A strategy for improving literature reviews in psychology courses. *Teaching of Psychology, 17,* 54–55.
Rickabaugh, C. A. (1993). The psychology portfolio: Promoting writing and critical thinking about psychology. *Teaching of Psychology, 20,* 170–172.
Russell, D. R. (1992). American origins of the writing-across-the-curriculum movement. In A. Herrington & C. Moran (Eds.), *Writing, teaching, and learning in the disciplines* (pp. 22–42). New York: The Modern Language Association of America.
Smith, P. C. (1995). Assessing writing and statistical competence in probability and statistics. *Teaching of Psychology, 22,* 49–50.

Sternberg, R. J. (1989). *The psychologist's companion: A guide to scientific writing for students and researchers.* New York: Cambridge University Press.

Stoloff, M. L., Apple, K. J., Barron, K. E., Reis-Bergan, M., & Sundre, D. (2004). Seven goals for effective program assessment. In D. S. Dunn, C. M. Mehrotra, & J. S. Halonen (Eds.), *Measuring up: Educational assessment challenges and practices for psychology* (pp. 29–46). Washington, DC: American Psychological Association.

Stoloff, M. L., & Rogers, S. (2002). Understanding psychology deeply through thinking, doing, and writing. *APS Observer, 15*(8), 21–22, 31–32.

Szuchman, L. T. (1999). *Writing with style: APA style made easy.* New York: Brooks/Cole.

Ware, M. E., Badura, A. S., & Davis, S. F. (2002). Using student scholarship to develop student research and writing skills. *Teaching of Psychology, 29,* 151–154.

Wilson, T. L., & Hershey, D. A. (1996). The research methods script. *Teaching of Psychology, 23,* 97–99.

About the Editors

Dana S. Dunn, a social psychologist, is Professor of Psychology and Director of the Learning in Common Curriculum at Moravian College, Bethlehem, PA. He received his PhD from the University of Virginia, having graduated previously with a BA in psychology from Carnegie Mellon University. A Fellow of the American Psychological Association, Dunn is active in the Society for the Teaching of Psychology. Dunn served as the Chair of Moravian's Department of Psychology from 1995 to 2001. He participated in the Psychology Partnerships Project in 1999. Dunn has written numerous articles, chapters, and book reviews concerning his areas of research interest: the teaching of psychology, social psychology, and rehabilitation psychology. Dunn is the author of three previous books—*The Practical Researcher, Statistics and Data Analysis for the Behavioral Sciences*, and *A Short Guide to Writing about Psychology*—and the co-editor of two others—*Measuring Up: Educational Assessment Challenges and Practices for Psychology* (with Chandra M. Mehrotra and Jane S. Halonen), and *Best Practices for Teaching Introduction to Psychology* (with Stephen L. Chew).

Randolph A. Smith. Randy Smith is Professor and Chair of the Department of Psychology at Kennesaw State University (Kennesaw, GA). He has taught at Kennesaw since 2003; prior to that time, he spent 26 years at Ouachita Baptist University (Arkadelphia, AR). Randy is a Fellow of the American Psychological Association (Divisions 1 and 2) and has filled a variety of positions within the Society for the Teaching of Psychology. Currently, he is Editor of *Teaching of Psychology*, a post he has held since 1997. He is co-author (with Steve Davis) of a research methods textbook, *The Psychologist as Detective: An Introduction to Conducting Research in Psychology*, and a combined statistics/research methods text, *An Introduction to Statistics and Research Methods: Becoming a Psychological Detective*. In addition, he has authored a critical thinking book, *Challenging Your Preconceptions: Thinking Critically About Psychology*, and has edited the Instructor's Manual for Wayne Weiten's introductory psychology text. Randy has more than 50 publications, including books, journal articles, and book chapters. In addition, he has given over 100 presentations and has supervised almost 150 undergraduate conference presentations. Randy's interests and research revolve around the scholarship of teaching of psychology. He was a co-founder of the Southwestern Conference for Teachers of Psychology and the Arkansas Symposium for Psychology Students, a student research conference that has existed for more than 20 years. He was a participant in the St. Mary's Conference in 1991 and served on the Steering Committee for the Psychology Partnerships Project in 1999. Randy is also a member of the Association for Psychological Science, Psi Chi, and the Southwestern Psychological Association (where he served as President in 1990–1991). In 2006, Randy received the Charles L. Brewer Distinguished Teaching of Psychology Award from the American Psychological Foundation. He earned his bachelor's degree from the University of Houston and his doctorate from Texas Tech University.

Bernard C. Beins. Barney Beins is Professor of Psychology and Chair of the Department at Ithaca College. He is a Fellow of APA, and was president of the Society for the Teaching of Psychology (2004) and its secretary from 1992 to 1994. He has taught at Ithaca College since 1986. Prior to that, he was at Thomas More College where he chaired the Psychology Department. He was also Director of Precollege and Undergraduate Education at APA (2000–2002). He is author of *Research Methods: A Tool for Life*, published by Allyn & Bacon, and he co-edited the *Gale Encyclopedia of Psychology*. He has about 80 publications, including books, journal articles, and book chapters. In addition, he has given over 150 presentations and has supervised about 70 undergraduate conference presentations. His interests and research revolve around the teaching of psychology. In 1994, he founded the Northeastern Conference for Teachers of Psychology, which continues on an annual basis in association with the New England Psychological Association convention. He also participated in the St. Mary's Conference in 1991 and in the Psychology Partnerships Project in 1999. He served as inaugural editor for the "Computers in Psychology" section of Teaching of Psychology from 1987 to 1996, and is currently an Associate Editor. He is also a member of the Association for Psychological Science, Sigma Xi, Psi Chi, the American Statistical Association, the Eastern Psychological Association, the New England Psychological Association, and the Rocky Mountain Psychological Association. He earned his bachelor's degree from Miami University in Oxford, Ohio, and his doctorate from City University of New York.

Author Index

I

J

K

L

Subject Index

S

Y

Z